Think Tank Media
Elmwood Park, New Jersey

Also by Sedrick D. Sims

Ancient African History: "A Journey Highlighting Africa's Past"

E VOLUTION

Sedrick Sims

Think Tank Media
Elmwood Park, New Jersey

This book is an original publication of Think Tank Media.

Copyright © 2013 by Sedrick Sims
Book cover design by Lola Jordan-Sims

First edition October 2014

Library of Congress Cataloging-in-Publication Data
Sims, Sedrick.
    Econvolution / Sedrick Sims.
ISBN 978-1-312-69035-6

This book is dedicated to Gabriel.

Introduction

Table of Contents

Introduction

The Oxford Dictionary defines *convoluted* as "especially of an argument, story, or sentence extremely complex and difficult to follow." The Oxford Dictionary also defines *convolution* as "coil or twist: the state of being or process of becoming coiled or twisted." These words summarize what I believe the theory of evolution through natural selection and Big Bang theories represent. After reading literature regarding the Big Bang and evolutionary theories, I have come to the conclusion that while these theories are good attempts at explaining life and world origins outside concepts of God and creationism, they have serious gaps in terms of explaining and understanding significant parts that make up these theories.

This book is written to express the opinion of a layperson that has read and studied scientific theory relating to evolution and the Big Bang. This book does not promote evolution, creation or intelligent design. I present issues discovered while researching evolutionary theory. In this book, I will highlight some of the details that make accepting evolutionary theory as fact *questionable*, that there are broad jumps, assumptions and conclusions made by some scientists, which makes some aspects of the theory of evolution unscientific.

Chapter 1
Charles Darwin, Science, Scientific Theory, Evolution?

Who was Charles Darwin? Charles Darwin was born on February 12, 1809, in Shrewsbury, England. His father, Robert Darwin, was a physician. Charles' mother, Susannah Wedgwood Darwin, died when he was just eight years old. Charles Darwin enrolled in Cambridge University hoping to become a clergyman in the Church of England. After graduating in 1831, Darwin accepted a job as a naturalist aboard the *HMS Beagle*. As a side note, the captain named Robert FitzRoy of the *HMS Beagle* was a depressed man who attempted suicide. He was concerned that he would not be able to complete the trip without being haunted by depression and suicidal thoughts. FitzRoy's uncle, Viscount Castlereagh, was a politician who had committed suicide. FitzRoy, even with his problems, was a supporter of slavery and human trafficking. Charles Darwin, unlike Captain Fitzroy, hated slavery. Before Fitzroy and Darwin took their trip on the *HMS Beagle,* the former Captain Pringle Stokes had committed suicide during the *Beagle's* first survey. It seems that the *HMS Beagle* had a troubling history. A few years after the trip with Charles Darwin, FitzRoy succeeded in killing himself. At 60 years old, he slit his throat with a razor [1].

Darwin's experience while traveling on the *HMS Beagle,* along with later experiments, were to be part of the foundation on which the theory of evolution was built. In 1859, Darwin published *On the Origin of Species*. Charles Darwin had to have been horrified by some of the offshoots from his theory, which would include the atrocities perpetuated under titles like "survival of the fittest" and "eugenics." The book *The Rough Guide to Evolution* mentions that Darwin's famous book did not deal with the origin of life, but he did write a letter to a Joseph Hooker that mentions the possibilities of life starting in a little pond with all sorts of chemicals – or abiogenesis. Abiogenesis is defined in the Oxford Dictionary as "the original evolution of life or living organisms from inorganic or inanimate substances"[2]. Regarding Darwin's letter to Joseph Hooker, according to Pallen quoting Darwin, "we have to assume that the first life forms originated under conditions that no longer exist" [3] . During his later years, Darwin was plagued by fatigue and problems with his intestines. He died on April 19, 1882, and was buried in Westminster Abbey. Pallen wrote, "Although the origin–of–life question remains unresolved, there has been considerable progress since Darwin's time. Instead of a blank sheet of paper, we now have the equivalent of a half–completed Sudoku puzzle, with some answers firmly written in ink, whereas in other cases several alternatives have been outlined with a pencil" [4].

Pallen's description is a good one of where we are in terms of our understanding of life origins. If you have ever played a Sudoku puzzle, depending upon its complication, it can be very hard to figure out even with "some answers written in ink." As a result, we have to test several assumptions in pencil to see if as we build our puzzle our assumptions still makes sense. Sometimes we can fill in several boxes and be almost complete only to find out that we have a wrong number in one of the boxes which makes the entire puzzle impossible to complete. We have to erase our assumption and rethink our conclusions. However, I think that wholly embracing evolutionary thought is quite different than playing a Sudoku puzzle. There is much more at stake which I will deal with in Chapter Five – what I believe to be some of the risks associated with this kind of view of life.

What is science?

The Oxford Dictionary defines science as, "the intellectual and practical activity encompassing the systematic study of the structure and behavior of the physical and natural world through observation and experiment"[5]. Some evolutionists believe that people who refuse to accept evolutionary theory hook line and sinker do not accept science at all, this is simply not true. When I write or speak about evolution and science, I always make it clear that we are grateful for what has been revealed to us through scientific experimentation and observation that has allowed us to improve our lives. We know there are things that science and nature has shown us to be true like gravity. We know that gravity exists. This is why when we awake we step out of the bed and we put our feet first, so that they can support us when we stand.

In the book *The Rough Guide to Evolution* by Mark Pallen, a PhD, we read "Evolution helps us to know our enemy, to understand – and fight against – changes in the biological world that threaten human health, wealth, happiness and even our very survival" [6]. I am not quite sure that I understand and agree that evolution helps us with our wealth. And I really doubt it can help us to be happy in life; actually, evolutionary thought could cause us to be the opposite – very sad.

Scientific Theory

What exactly is a scientific theory? The Encyclopedia Britannica defines scientific theory as "systematic ideational structure of broad scope, conceived by the human imagination, that encompasses a family of empirical (experiential) laws regarding regularities existing in objects and events, both observed and posited. A scientific theory is a structure suggested by these laws and is devised to explain them in a rational manner. In attempting to explain things and events, the scientist employs: (1) careful observation or experiments; (2) reports of regularities; and (3) systematic explanatory schemes (theories)" [7] . The website about.com defines scientific theory as "...an idea starts out as a hypothesis, or an educated guess as to why or how something is happening. Those hypotheses undergo a lot of experimental trials in various ways and data is collected. Over time, as more and more data is collected, a hypothesis can become a scientific theory. The more a hypothesis is supported by many different experiments, the stronger the scientific theory"[8]. Scientific theories always have the potential to be overturned. Karen Fox wrote in *The Big Bang What it is, Where it Came from and Why it Works* "They are never proven beyond a shadow of a doubt. Just as believing dogmatically in one version of any myth or religion leads only to narrow understanding of the world, so dogmatically believing in any scientific theory is problematic" [9].

The popular media often present theories that scientists come up with to us the general public minus the supporting details. These details could cause us to reject some of these theories. Some of us believe that scientists are open-minded and would surely change their views if there was significant evidence that proved their theories mostly incorrect. Karen Fox wrote that the general public thinks that scientists are free, open-minded persons that would adjust or change their theories once they saw that their ideas were wrong. "The way many people like to imagine the process is that everyone keeps and admirably open mind and changes his or her beliefs when enough contradictory proof collapses an old theory"[10]. Fox quoted a Harvard University science professor who said scientists worked within their theories long after they had factual proof to destroy them. These scientists did not discard their theories until it had been totally destroyed by new facts or theories. This should be kept at the forefront of our minds when reviewing scientific theories that are disseminated to the public. The general public does not understand that scientific theories are handled this way, so how can we properly decipher that which is true and more up to date if the scientists themselves are unwilling to change and will hold onto theories that are outdated.

What is Evolution?

The PBS.org website states that biological evolution "…refers to the cumulative changes that occur in a population over time. These changes are produced at the genetic level as the organisms' genes mutate and or recombine in different ways during reproduction and are passed on to future generations. Sometimes, individuals inherit new characteristics that give them a survival and reproductive advantage in their local environments; these characteristics tend to increase in frequency in the population, while those that are disadvantageous decrease in frequency. This process of differential survival and reproduction is known as natural selection. Non-genetic changes that occur during an organism's life span, such as increases in muscle mass due to exercise and diet, cannot be passed on to the next generation and are not examples of evolution"[11]. The University of California Museum of Paleontology website, supported by the National Science Foundation and the Howard Hughes Medical Institute, defines biological evolution as "…descent with modification. This definition encompasses small-scale evolution (changes in gene frequency in a population from one generation to the next) and large-scale evolution (the descent of different species from a common ancestor over many generations). Evolution helps us to understand the history of life"[12].

One of the ways evolutionists create confusion about evolution is when they assume that microevolution is proof that macroevolution occurs. I do not think this generalization is by accident but is intentionally done to gloss over and hide the truth. When macroevolution is challenged, microevolutionary proof is twisted with the assumption that macroevolution is the same. "Microevolution happens on a small scale (within a single population), while macroevolution happens on a scale that transcends the boundaries of a single species. Despite their differences, evolution at both of these levels relies on the same, established mechanisms of evolutionary change"[13]. Most of the examples that are used today by the media and evolutionists are of microevolution within a single population affecting that specific species. Macroevolution is where we have big problems. To think that distinctly different species would develop from a common ancestor over many, many years is incredible, especially without solid, verified examples of macroevolution. Nature.com posted a story about Darwin, microevolution and macroevolution, noting that "Macro evolution posed a problem to Darwin because his principle of descent with modification predicts gradual transitions between small scale adaptive changes in populations and these larger-scale phenomena, yet there is little evidence for such transitions in nature. Instead, the natural world is often characterized by gaps, or discontinuities. One type of gap relates to the existence of 'organs of extreme perfection', such as the eye, or morphological innovations, such as wings, both of which are found fully formed in present-day organisms without leaving evidence of how they evolved" [14]. The paper *The Cambrian Explosion: Biology's Big Bang* adds, "Most biologists now acknowledge that the Darwinian mechanism of natural selection acting on random variations can explain small-scale micro-evolutionary changes, such as cyclical variations in the size of the beaks of Galapagos finches or reversible changes in the expression of genes controlling color in English peppered moths" [15].

Scientists have studied lifeforms for years and are aware of the hint of intelligence shown in them. In 1994, Francisco Ayala, as president of the American Association for the Advancement of Science, wrote, "The functional design of organisms and their features would therefore seem to argue for the existence of a designer. It was Darwin's greatest accomplishment to show that the directive organization of living beings can be explained as the result of a natural process, natural selection, without any need to resort to a Creator or other external agent"[16]. As we will see later, this great accomplishment regarding living organisms is a good attempt at theorizing life as we know it, which is partially based upon a few scientific facts. However, there are many questions that are unanswered and parts of the theory that gloss over the hard questions.

NOTES

1. Pallen, MJ. The Rough Guide to Evolution. Rough Guides. London; New York: Penguin Group 2009. 25 p.

2. Oxford Dictionary online. Abiogenesis [Internet]. 2014 Jan 14: Oxford University Press; Available from http://www.oxforddictionaries.com/definition/english/abiog enisis

3. Pallen, MJ. The Rough Guide to Evolution. Rough Guides. London; New York: Penguin Group 2009. 149 p.

4. Pallen, MJ. The Rough Guide to Evolution. Rough Guides. London; New York: Penguin Group 2009. 149–150 p.

5. Oxford Dictionary online. Science [Internet]. 2014 Jan 14: Oxford University Press; Available from http://www.oxford dictionaries.com/definition/english/science

6. Pallen, MJ. The Rough Guide to Evolution. Rough Guides. London; New York: Penguin Group 2009. iv–v p.

7. *Encyclopædia Britannica Online.* Scientific theory [Internet]. [2014, Sep 23]. Available from http://www.britannica.com /EBchecked/topic/528971/scientific-theory

8. Scoville, H. What is a scientific theory? About.com [Internet]. [2014 Sep 23]. Available from http://evolution.about.com /od/Overview/g/What-Is-A-Scientific-Theory.htm

9. Fox, KC. The Big Theory– What It Is, Where It Came From. New York (NY): John Wiley & Sons, Inc.; 2002. 5 p.

10. Fox, KC. The Big Theory– What It Is, Where It Came From. New York (NY): John Wiley & Sons, Inc. 2002. 8 p.

11. WGBH Educational Foundation, Clear Blue Sky Productions, Inc. Biological Evolution. PBS.org [Internet]. [2014 Sep 23]. Available from http://www.pbs.org /wgbh/evolution /library/faq/cat01.html

12. Understanding Evolution Team. Understanding Evolution – Your one-stop source for information on evolution [Internet]. [2014 Sep 23]. Available from http://evolution.berkeley.edu /evolibrary/article/evoscales_01

13. Understanding Evolution Team. Understanding Evolution – Your one-stop source for information on evolution [Internet]. [2014 Sep 23]. Available from http://evolution.berkeley. edu/evolibrary/article/evoscales_01

14. Reznick DN, Ricklefs RE. Darwin's bridge between microevolution and macroevolution. Nature. 2009 Feb 12;457(12): 837-840 doi10.1038/nature07894

15. Meyer SC, Nelson, PA, Chien P. The Cambrian Explosion: Biology's Big Bang [Internet]. Discovery Institute; [2014 Sep 23]. Available from http://www.discovery.org/article/Files /PDFs/Cambrian.pdf

16. Ayala, F. Darwin's Revolution. Boston (MA): *Creative Evolution?* Jones and Bartlett Publishers; 1994. 4-5 p.

Chapter 2
What is Cosmology?

Cosmology, according to Joseph Silk, "is the study of large-scale structure and evolution of the universe" [1]. Silk, in his book *The Big Bang, the Creation and Evolution of the Universe,* wrote, "Although we are not able to answer all the central questions of cosmology, the Big Bang theory provides us with a broad outline of the evolution of the universe"[2]. The Big Bang theory is a theory of the origin of the universe. The theory states that the universe began from an initial denser and hotter point and then expanded over billions of years to form the universe today. Silk wrote, "The early universe was very hot, very dense, and perhaps also very irregular. The irregularity and anisotropy gradually decayed. Within minutes after the Big Bang, some nuclear reactions occurred; essentially all the helium in the universe was synthesized at that time. As the universe expanded, it cooled, much as hot air expands and cools. The cosmic background radiation is a residual vestige of this early era; it has been aptly christened the relic radiation of the *Primeval Fireball.* As the matter in the universe cooled, it eventually condensed into galaxies, according to one scenario for the evolution of the universe. The galaxies fragmented into stars and clustered together to form great aggregations over vast regions of space. As the first generation of stars were born and died, the heavy elements, such as carbon, oxygen, silicon, and iron, were gradually synthesized"[3]. I often say "sounds real, real good" when I hear or read things that are so neatly arranged and laid out for my consumption. All of this reads very clear and simple. However, when we begin to ask who, what, when, where, why and how the questions that if answered gives us a complete story, things start to get real fuzzy and less certain.

When we look at the details, we begin to see the weakness in these theories. Fox wrote, "The minutiae matter in science. It is the minutiae that disprove a false theory and support a true one"[4].

From Albert Einstein to Russian meteorologist Aleksander Friedman, both have created models of the universe. Friedmann's models were the first to conclude that there was a start of time. A good question would be why did time start? What was before time? Did nothingness somehow wind itself up like a clock to start time? It is interesting that from Charles Darwin to George Lemaiter, we have religious persons offering explanations for the universe's creation. George Lemaiter, a priest in Belgium, created a Big Bang model. Both Darwin and Lemaiter came up with a beginning point in space and time for everything, when the entirety of existence was combined into a single spot.[5] Of course, from this single spot the whole universe sprung. The Big Bang theory states that the universe was much hotter and denser at the start than it is now. If this was true, it would be in conflict with the cosmological principle that states that the mass in the universe must be same at all times. The Big Bang theory here is assisted by intertwining additional theories to "fix" and allow continued acceptance of it. Scientists then came up with the Steady State theory, which the Oxford dictionary defines as, "…an unvarying condition in a physical process, especially as in the theory that the universe is eternal and maintained by constant creation of matter." The merging of this new theory brought the Big Bang theory inline with cosmological theory. In 1948, Herman Bondi, Thomas Gold, Fred Hoyle wrote about the universe, "This theory predicts a similar aspect for the universe at all times by postulating that matter is being continuously created at precisely the right rate to maintain the same mean matter density everywhere in the universe"[6]. In the book *The Big Bang,* Silk wrote concerning the time before and right up to the Big Bang that "Conditions at this initial instant and before this instant are matters for speculation that the conventional theory does not address"[7] .

Silk noted in his preface that "The bulk of the Big Bang theory, however rests more on fact than on speculation"[8]. This statement lets us know that the Big Bang theory, according to Silk, consists of both fact and speculation. However, the theory's foundation is mostly based upon facts as opposed to the speculative parts. When I think about scientific theories and the definition I do not have in the back of my mind that some parts of the theory will be speculative. What is speculation? The Oxford Dictionary defines it as "the forming of a theory or conjecture without firm evidence"[9]. The general public should know what part of the "scientific theory" is factual and what part is speculation. This has been one of my main objections when scientific theories are presented to the public. Fox mentions that the Big Bang theory is universally accepted today because it is taught as fact. The Big Bang theory needs to the coupled with the details so that the public will be able to decide whether or not the theory is solid. Stick to what we know not what we are assuming. Fox wrote, "The majority of the world population that accepts the Big Bang theory does so unquestioningly"[10]. This is why I felt the need to do my part in highlighting the questions, dilemmas and ideas that are simply unclear when it comes down to evolutionary thought. Accepting ideas as facts without questioning is not wise. Instead, ask questions like "How do we know that this is what happened, do we have any direct solid evidence?" In a courtroom, a witness cannot be considered a reliable witness if she/he reveals that they did not actually witness the crime or law that was broken by individual(s). Even the Bible tells us to examine what we are taught in Acts 17:11 to see if they are true. Joseph Silk mentions the lack of evidence regarding the origins of the universe, "But we would like to find some more direct evidence from this early era of the universe. Of course, we cannot directly observe the Primeval Fireball, and in fact direct observation would have been impossible even by a hypothetical human observer, for the universe did not become transparent until after one million years"[11]. Remember, we are discussing science here not religious faith which we cannot possibly prove. Somehow, we are expected to deny faith in God and swallow all evolutionary theory as fact even though no one has and ever will have firsthand experience. We are expected to accept cosmology along with evolution as fact even though no one can provide the details. We are further confused by the fact that we cannot prove significant parts of the theories because it is impossible to test.

A scientist, George Gamow was a professor at George Washington University. He studied with Aleksander Friedmann in Russia and Ernest Rutherford, the discoverer of the atomic nucleus. Ernest Rutherford understood that the atoms that exist today would have to have been created in the first explosion. George Gamow wanted to understand why we have the elements in their current amounts. He suggested that the results of the big explosion could be measured and detected today. George Gamow also predicted the variety of temperature of the radiation from this explosion. Arno Penzias and Robert Wilson were the first scientists credited with discovering/hearing the distant hum of radiation in the background predicted by Gamow. These two were actually working on a Bell Labs project where they used an old antenna for radio astronomy. On June 1964 they discovered radiation at 7.35 centimeter wavelength. They thought that they should have had silence, which proved their instruments were adjusted properly. Fox wrote "The work is generally heralded as the first real proof to support the big bang (and to this day, the existence of the cosmic microwave background is one of the most important pieces of evidence). It is now generally agreed that the universe evolved over time, starting from some initial form of radiation and particles. The Big Bang theory had finally entered the world of 'real' science"[12].

The Big Bang itself is amazingly precise and structured because if the rate that the universe expanded was different by a small amount the galaxies would have never formed with such accuracy. Yes, what accuracy! Why would the pinpoint of energy expand only thus far so that the galaxies would form the way they did? How did this pinpoint of energy happen to develop the way it did and how did it know when to reign in its power to prevent self-implosion/destruction. Fox wrote concerning the Big Bang, "If the strength of the nuclear weak force was different, all hydrogen in the universe would have been burned into helium, and there would be no water"[13]. Even within the idea of a Big Bang, there is a hint of precision radiating from the forming of the world and universe, a whisper of an organized formation. Fox mentions that scientists tried to explain how a random energy fluctuation created such precision but to no avail [14]. I always have wondered where did this pinpoint of energy come from and why did it begin as a small dense powerful energy.

In the book *Conflict in the Cosmos Fred Hoyle's Life in Science* by Simon Milton, there is part of a lecture that Fred Hoyle gave at the Royal Astronomical Society (RAS) in 1973. Fred Hoyle was a popularizer of science and he entertained his audience with these words, "The discovery of the expansion of the system of galaxies poses a problem with the origin of the universe which after fifty years of work by cosmologists appears perhaps more mysterious today than it did at the end of the first quarter of the century. Much effort has been expended in discussing alternative cosmological models, with the hope that a particular model might be singled out through a better agreement with observation. This effort has been unproductive, however, since even five decades of accumulated data have not sufficed to distinguish with sufficient accuracy between the various possibilities"[15]. Hoyle explained that there were conceptual complexities when we try to use the Big Bang to explain life origins that are "so insuperable that it seemed almost inevitable that one had to by-pass them"[16] . Hoyle was referring to the need for the steady-state model that involves the creation of matter. The leading cosmological theory of life origins had to be assisted with the steady state model so that it remained tenable. Hoyle also said in the speech that the observable universe is not by itself but a part of a larger unobservable universe. In other words, what we call the universe is part of a larger whole[17].

Returning to cosmic radiation, we see that the radiation is uniform and the same from every direction. So within one second after the explosion, the universe is created smooth and evenly heated. This is why the universe is the same today. When I think of a Big Bang or explosion, I see utter chaos and things being strewn everywhere at different speeds and angles. Whatever was in place before the explosion would now be garbled and mixed together. As Fox mentions, why then were galaxies formed if everything was smooth and evenly heated there would be nothing to get caught together by force or gravity to create huge galaxies. Writing about the early universe, Fox states, "The universe, small even though it was, didn't have enough time to reach a uniform temperature. So somehow it must have been smooth from the get-go, instead of having a chance to grow smooth with time. The question is why. This is known as the horizon problem"[18]. Please tell me how does nothing but locked-up energy and void get described as smooth when there was nothing but a small pinpoint somewhere in the midst of nothingness? How do we get smoothness unless everything at this point is rough or bumpy?

Another mystery connected to the Big Bang theory is dark matter. Dark matter is defined in some cosmological theories non-luminous material, which is postulated to exist in space and which could take either of two forms: weakly interacting particles (cold dark matter) or high-energy randomly moving particles created soon after the Big Bang (hot dark matter). Fox wrote, "Most astronomers take the existence of dark matter for granted, but actually hammering down what it is hasn't been successful (and no one has even been able to find as much of it as one would like to see; ..)"[19]. Dark matter is necessary to prove that the universe is at a critical density. Critical density occurs when the universe has enough mass/volume to close or collapse the universe but too little mass/volume to keep from expanding. Alexander Friedmann of Russia developed an equation for the expanding universe in the 1920s. The term critical density is the average density of matter required for the Universe to just halt its expansion, but only after an infinite time. This results in a universe that would have to be flat. Fox wrote, "Historically, there has been a certain glossing over of many of the questions the big bang theory has brought up...Dark matter offers hope that the universe is at a critical density, but it's mostly math and gravity – no one has completely hammered out the particulars yet"[20]. As I reviewed reports on dark matter I, came across *Origins: Fourteen Billion Years of Cosmic Evolution* by Tyson and Goldsmith, who wrote about the missing mass problem which is mass that is expected to be within galaxies is not visible or discernible. Tyson and Goldsmith wrote that this problem was discovered in 1933 and then in 1976 Vera Rubin found the same problem. Rubin discovered that inside of the visible disk of the galaxy that the stars farther away from the center moved at faster speeds than the closer stars. This correlates because the stars furthest away have the greater matter between the galaxy and themselves. This would require higher speed to stay in orbit. Since this dark matter is six times as available as ordinary matter, Tyson and Goldsmith believes that we should be able to find "that one in every six pieces of dark matter has a chunk of ordinary matter clinging to it"[21]. If this was true, we should be able to identify dark matter by the visible matter stuck to it.

Alex Stone wrote in the periodical *Dark Matter Made Visible* that physicists suspect that 80 percent of all matter is "dark," which means it does not emit light and does not interact with normal matter except through gravity. Stone wrote, "Or that's what physicists have long suspected, without direct proof" [22]. Alex Stone mentions the colliding galaxy clusters in the constellation Carina that were studied by Marusa Bradac at Stanford University. The conclusion was that dark matter did exist. Alex Stone wrote, "The best evidence for dark matter was that orbital speeds of stars in a galaxy do not fall off with increasing distance from the galaxy's center, as would seem to be necessary to keep the stars from flying off into space. The fact that the galaxies hold together suggests that unseen mass provides the gravity to hold them together" [23].

Albert Einstein gave us the formula $E=mc^2$, or energy equals mass times the square of the speed of light. Tyson and Goldsmith wrote, "With this equation, you can tell how much radiant energy a star can produce, or how much you could gain by converting the coins in your pocket into useful form of energy"[24]. They add that "dark matter has about six times the mass of all the visible matter"[25] and "Thus as best we can figure, the dark matter doesn't simply consist of matter that happens to be dark. Instead, it's something else altogether. Dark matter exerts gravity according to the same rules that ordinary matter follows, but it does little else that might allow us to detect it. Of course we are hamstrung in this analysis by not knowing what the dark matter is"[26]. In the *Prospects for detecting supersymmetric dark matter in the Galactic halo*, a journal article published in Nature on November 1, 2008, we find "Dark matter is the dominant form of matter in the Universe, but its nature is unknown"[27]. Another interesting report, *Dark matter and dark energy,* by Robert Caldwell and Mark Kamionkowski, stated that "Observations continue to indicate that the Universe is dominated by invisible components – dark matter and dark energy"[28]. Caldwell and Kamionkowski explain that the composition of the universe is directly related to the Big Bang model. "In terms of their contribution to the mean energy density, the contents of the Universe are approximately 75% dark energy, 20% dark matter and 5% normal (atomic) matter, with smaller contributions from photons and neutrinos. These measurements rely on the validity of the hot Big Bang model, general relativity and the cosmological principle (that the Universe is uniform on the largest scales)" [29] . We see that this dark matter idea is very important and is closely linked to cosmology and assumptions about the universe. Even though we cannot comprehend dark matter, Caldwell and Kamionkowski wrote "We can infer the presence of dark matter through indirect methods, despite not being able to see it"[30]. The authors admit that dark matter does not coincide with our current scientific understanding and state, "The breadth and depth of experiments and observations that support these under lying tenets give us confidence that this model of the cosmos has a solid foundation"[31]. The jury is still out our understanding of dark matter is minimal, yet cosmologists and evolutionists have already reached their verdict despite the lack of understanding of the evidence so far provided?

Tyson and Goldsmith wrote that astrophysicists are uncomfortable with the dark matter concept: "So, dark matter is our friend. But astrophysicists understandably grow uncomfortable whenever they must base their calculations on concepts they don't understand, even though this wouldn't be the first time they've done so"[32]. This is revealing in that the acceptance of ideas without understanding is not exactly foreign to the scientist. At what point do we the public, not knowing the details behind the theory, decipher when to accept the theory? Tyson and Goldsmith wrote, "We're not inventing dark matter out of thin space; instead, we deduce its existence from observational facts[33]. A seeing–is–believing approach to life works well in many endeavors, including mechanical engineering, fishing and perhaps dating, but it doesn't make for good science. Science is not just about seeing. Science is about measuring – preferably with something that's *not* your own eyes, which are inextricably conjoined with the baggage of your brain: preconceived ideas, post–conceived notions, imagination unchecked by reference to other data, and bias"[34]. Tyson and Goldsmith later in their book wrote about the origin of life on Earth. They wrote that Darwin's letter to Joseph Hooker, in which Darwin suggested that life began in a little pond, and later Stanley Miller– Harold Urey 1953 tested Darwin's idea. Tyson and Goldsmith wrote, "From a scientific perspective, nothing succeeds like experiments that can be compared with reality."[35]. Hold up, is this a reversal of the importance of experimental seeing–is–believing? Darwin used what he saw and viewed with his own eyes to formulate his theories. Stanley Miller and Harold Urey in 1953 were able to see that chemicals could combine in an assumed simulated prebiotic earth. This type of back and forth accept–proof–now–reject–proof–later process is repeated by Richard Dawkins in his book *Evolution the Greatest Show on Earth* about the incomplete fossil record, when he writes, "The fossil record…is a bonus, something that we had no right to expect as a matter of entitlement. There is more than enough evidence for the fact of evolution in the comparative study of modern species (chapter 10) and their geographical distribution (chapter 9)"[36]. Dawkins then states, "We don't need fossils – the case for evolution is watertight without them; so it is paradoxical to use gaps in the fossil record as though they were evidence against evolution. We are, as I say, lucky to have fossils at all"[37].

Tyson and Goldsmith raised an interesting question about dark matter: "If all matter has mass, and all mass has gravity, does all gravity have matter?" [38]. They add, "More likely, dark matter consists of matter whose nature we have yet to divine, and which clusters more diffusely than ordinary matter does"[39]. Move over periodic table or law of physics because a new sheriff is in town dark matter. This thing defies everything we know about matter, it will be interesting if we learn exactly what it is. This mystery of dark matter baffles the mind especially of astrophysicists. Tyson and Goldsmith wrote, "But what if this conclusion were entirely wrong? When nothing else seems to work, some scientists will understandably, and quite rightly, question the fundamental laws of physics that underlies the assumptions made by others who seek to understand the universe"[40]. The Big Bang is the most widely accepted theory about the beginning of the universe. There are others theories that are not as popular. However, the moment a new theory arises and becomes dominant, Fox wrote "...these theories will be accepted as wholeheartedly, and as quickly, as the big bang. In much the same way, it will not be the clinching experiments that give the proof of its robustness, but the way the theory makes sense in the minds of those who hear it"[41]. It is amazing that this is the way some scientists deal with scientific theories.

Tyson and Goldsmith also wrote that the origin of life is locked in murky uncertainty. This is because our knowledge of the events that occurred billions of years ago cannot be directly examined. Tyson wrote, "For times more than 4 billion years in the past, the fossil and geological record of Earth's history does not exist" [42]. In fact, Tyson wrote, "The absence of all geological evidence from the epochs more than 4 billion years ago arises from motions of Earth's crust, familiarly called continental drift but scientifically known as plate tectonics"[43]. Plate tectonic motion buried everything that laid on Earth's surface. Because of this, we have very few rocks that date older than 2 billion years of course and none that are more than 3.8 billion years old.

Francis Crick mentions that discussions of the origin of the universe focuses around the idea of a big bang. This idea "is built up from our present–day knowledge of the fundamental particles of matter and radiation, together with a rather small number of experimental facts, such as the cosmic radiation background which now pervades all space – the faint whisper of creation just audible in radio telescopes. Such an imaginative synthesis is necessarily not entirely secure. Weinberg confesses to an occasional feeling of unreality in writing about it" [44].

Crick mentions that when looking at the Big Bang theory we have three significant pieces of evidence: 1. The universe is expanding; 2. The universe has the right amount of helium and other atoms; 3. We think we've found the radiation leftover from initial explosion. This evidence is not what I would consider irrefutable.

Tyson and Goldsmith wrote about before and immediately after the Big Bang, "But what happened before all this cosmic fury? What happened before the beginning? Astrophysicists have no idea. Rather, our most creative ideas have little or no grounding in experimental science."[45] What do we have so far in terms of our knowledge of the beginning of everything is that before the dense intense pinpoint of immense power we know nothing. This knowing nothing includes how this powerful minute pinpoint came to be in the first place. Why this explosion happened we do not know. How long this enormously powerful pinpoint existed we do not know. Why this powerful source of power evolved as it did we do not know. What we claim to know is that the pinpoint of enormous power exploded/expanded and supposedly created the universe. This universe, planets, and stars eventually aided abiogenesis. I wonder with so many unknowns especially from the earliest point that we know nothing at all about, how can we be so sure about all that follows?

NOTES

1. Silk, J. The Big Bang : The Creation and Evolution of the Universe. Berkley (CA): W. H. Freeman and Company; 1980. 1 p.

2. Silk, J. The Big Bang : The Creation and Evolution of the Universe. Berkley (CA): W. H. Freeman and Company; 1980. 2 p.

3. Silk, J. The Big Bang : The Creation and Evolution of the Universe. Berkley (CA): W. H. Freeman and Company; 1980. 6 p.

4. Fox, KC. The Big Theory- What It Is, Where It Came From. New York (NY): John Wiley & Sons, Inc. 2002. 25 p.

5. Fox, KC. The Big Theory- What It Is, Where It Came From. New York (NY): John Wiley & Sons, Inc. 2002. 46 p.

6. Silk, J. The Big Bang : The Creation and Evolution of the Universe. Berkley (CA): W. H. Freeman and Company; 1980. 4 p.

7. Silk, J. The Big Bang : The Creation and Evolution of the Universe. Berkley (CA): W. H. Freeman and Company; 1980. 6 p.

8. Silk, J. The Big Bang : The Creation and Evolution of the Universe. Berkley (CA): W. H. Freeman and Company; 1980. xiv p.

9. Oxford Dictionary online. Speculation [Internet]. 2014 Jan 14: Oxford University Press; Available from http://www.oxford dictionaries.com/definition/english/speculation

10. Fox, KC. The Big Theory- What It Is, Where It Came From. New York (NY): John Wiley & Sons, Inc. 2002. 119 p.

11. Silk, J. The Big Bang : The Creation and Evolution of the Universe. Berkley (CA): W. H. Freeman and Company; 1980. 127 p.

12. Fox, KC. The Big Theory– What It Is, Where It Came From. New York (NY): John Wiley & Sons, Inc. 2002. 80 p.

13. Fox, KC. The Big Theory– What It Is, Where It Came From. New York (NY): John Wiley & Sons, Inc. 2002. 85 p.

14. Fox, KC. The Big Theory– What It Is, Where It Came From. New York (NY): John Wiley & Sons, Inc. 2002. 86 p.

15. Mitton, S. Conflict In The Cosmos Fred Hoyle's Life In Science. Washington, (DC): Joseph Henry Press, 2003. 318 p.

16. Mitton, S. Conflict In The Cosmos Fred Hoyle's Life In Science. Washington, (DC): Joseph Henry Press, 2003. 319 p.

17. Mitton, S. Conflict In The Cosmos Fred Hoyle's Life In Science. Washington, (DC): Joseph Henry Press, 2003. 319 p.

18. Fox, KC. The Big Theory– What It Is, Where It Came From. New York (NY): John Wiley & Sons, Inc. 2002. 99 p.

19. Fox, KC. The Big Theory– What It Is, Where It Came From. New York (NY): John Wiley & Sons, Inc. 2002. 107 p.

20. Fox, KC. The Big Theory– What It Is, Where It Came From. New York (NY): John Wiley & Sons, Inc. 2002. 116 p.

21. Tyson, ND, Goldsmith, D. Origins Fourteen Billion Years of Cosmic Evolution. New York–London: W.W. Norton & Company; 2005. 70 p.

22. Stone, A. 2006 Nov 13. Dark matter made visible. Discover, 27(11): 18 p.

23. Stone, A. 2006 Nov 13. Dark matter made visible. Discover, 27(11): 18 p.

24. Tyson, ND, Goldsmith, D. Origins fourteen billion years of cosmic evolution. New York–London: W.W. Norton &

Company; 2005. 36 p.

25. Tyson, ND, Goldsmith, D. Origins fourteen billion years of cosmic evolution. New York–London: W.W. Norton & Company; 2005. 68 p.

26. Tyson, ND, Goldsmith, D. Origins fourteen billion years of cosmic evolution. New York–London: W.W. Norton & Company; 2005. 69 p.

27. Springel, V, White, SDM, Frenk, CS, Navarro, JF, Jenkins, A, Vogelsberger, J, Wang, A. Ludlow, A, Helmi, A Prospects for detecting supersymmetric dark matter in the Galactic halo. 2008 Nov 6; 456: 73-76.

28. Robert C, Kamionkowski, M. Nature dark matter and dark energy. 2009 Apr 2; 458: 587-589.

29. Robert C, Kamionkowski, M. Nature dark matter and dark energy. 2009 Apr 2; 458: 587-589.

30. Robert C, Kamionkowski, M. Nature dark matter and dark energy. 2009 Apr 2; 458: 587-589.

31. Robert C, Kamionkowski, M. Nature dark matter and dark energy. 2009 Apr 2; 458: 587-589.

32. Tyson, ND, Goldsmith, D. Origins fourteen billion years of cosmic evolution. New York–London: W.W. Norton & Company; 2005. 75 p.

33. Tyson, ND, Goldsmith, D. Origins fourteen billion years of cosmic evolution. New York–London: W.W. Norton & Company; 2005. 74 p.

34. Tyson, ND, Goldsmith, D. Origins fourteen billion years of cosmic evolution. New York–London: W.W. Norton & Company; 2005. 74 p.

35. Tyson, ND, Goldsmith, D. Origins fourteen billion years of cosmic evolution. New York–London: W.W. Norton & Company; 2005. 224 p.

36. Dawkins, R. The Greatest Show on Earth: The Evidence for Evolution. New York, (NY): Free Press; 2009. 146 p.

37. Dawkins, R. The Greatest Show on Earth: The Evidence for Evolution. New York, (NY): Free Press; 2009. 146 p.

38. Tyson, ND, Goldsmith, D. Origins fourteen billion years of cosmic evolution. New York–London: W.W. Norton & Company; 2005. 69 p.

39. Tyson, ND, Goldsmith, D. Origins fourteen billion years of cosmic evolution. New York–London: W.W. Norton & Company; 2005. 70 p.

40. Tyson, ND, Goldsmith, D. Origins fourteen billion years of cosmic evolution. New York–London: W.W. Norton & Company; 2005. 70 p.

41. Tyson, Neil DeGrasse and Goldsmith, Donald *Origins Fourteen Billion Years of Cosmic Evolution* – W.W. Norton & Company New York–London 2005. 70 p.

42. Fox, KC. The Big Theory– What It Is, Where It Came From. New York (NY): John Wiley & Sons, Inc. 2002. 117 p.

43. Tyson, ND, Goldsmith, D. Origins fourteen billion years of cosmic evolution. New York–London: W.W. Norton & Company; 2005. 256 p.

44. Tyson, ND, Goldsmith, D. Origins fourteen billion years of cosmic evolution. New York–London: W.W. Norton & Company; 2005. 256 p.

45. Crick, F. Life itself : its origin and nature. New York (NY): Simon and Schuster; 1981. 30 p.

46. Tyson, ND, Goldsmith, D. Origins fourteen billion years of

cosmic evolution. New York–London: W.W. Norton & Company; 2005. 44 p.

Chapter 3
Probability

     Probability is used often in science and life every day. The website Cancer.org states that the "The lifetime risk of developing or dying from cancer refers to the chance a person has, over the course of his or her lifetime (from birth to death), of being diagnosed with or dying from cancer…The risk is expressed both in terms of a percentage and as odds. For example, the risk that a man will develop bladder cancer during his lifetime is 3.82%. This means he has about 1 chance in 26 of developing bladder cancer (100/3.82 = 26). Put another way, 1 out of every 26 men in the United States will develop bladder cancer during his lifetime"[1]. The website www.mayoclinic.com notes, "You might wonder what your chances are of developing cancer. But cancer statistics can be confusing. News reports make it sound as if nearly every day something is found to dramatically raise your risk. Sorting through all the information and figuring out what's valid and what isn't can be tricky. When scientists talk about risk, they're referring to a probability – the chance that something may occur, but not a guarantee that it will. For example, if you flip a coin, there is a one in two chance, or a 50 percent chance, that the coin will land heads up"[2].

     When we try to apply probability or chance that life began as explained through evolution, we cannot get away from the serious problems with just trying to set up the formula. Probability is the chance or likelihood that an event could happen. The ratio of the number of ways and event can happen to the number of possible outcomes.

(P)   Probability = of ways a certain outcome can occur

Total Possible outcome. (sample space)

A simple example of probability using a standard die with six sides. The die has a number of dots from 1 dot to 6 dots. The question what is the probability of getting 5 dots if we roll the die (P5). We know that there is only 1 area on a die with 5 dots. There are 6 possible outcomes (sample space – used in denominator) when rolling 1 die. Therefore our formula for the probability of rolling a 5 when we roll one die is:

P (5) = 1   (of ways a certain outcome can occur)

----

6   [Total Possible outcome. (sample space)]

The lottery has a probability of picking only 1 set of the 6 correct numbers is about 1 in 14,000,000, "Therefore, the odds of any given set of numbers coming up in this type of lottery are roughly 1 in 14 million"[3]. Many people will not play the lottery simply because they understand that their chances or probability of winning is very low. Therefore, they do not buy into the idea of "being in it to win it."

With evolution, one can reasonably expect that the unknown probability for the birth of the first cell (abiogenesis) is very, very, very small [3]. The website Answersingenesis.org gives a loose, generous probability of a stable gene being formed by itself but notes, "Nevertheless, let us use as many sets as there are atoms in the universe. Let us give chance the unbelievable number of attempts of eight trillion tries per second in each set! At this speed on average, it would take 10^{147} years to obtain just one stable gene"[4].

Frances Crick mentions the frustrations when trying to work with the series of events and their complexities and probabilities relating to the evolutionary theory of universe and life origins when he wrote, "What is so frustrating for our purpose is that it seems almost impossible to give *any* numerical value to the probability of what seems a rather unlikely sequence of events"[5]. Loosely calculating probability, Crick wrote, "However if p was only one chance in a billion, billion, the chance of starting was not far from even. If as little as one in 10^{15} (A thousand billion, billion), the chance of life starting here was very small"[6]. In the article *Probability and Order Versus Evolution*, Henry Morris, Ph.D. Morris wrote. "Astro–physicists estimate that there are no more than 10^{80} infinitesimal 'particles' in the universe, and that the age of the universe in its present form is no greater than 10^{18} seconds (30 billion years). Assuming each particle can participate in a thousand, billion (10^{12}) different events every second (this is impossibly high, of course), then the greatest number of events that could ever happen (or trials that could ever be made) in all the universe throughout its entire history is only 10^{80} x 10^{18} x 10^{12}, or 10^{110} (most authorities would make this figure much lower, about 10^{50}). Any event with a probability of less than one chance in 10^{110}, therefore, cannot occur. Its probability becomes zero, at least in our known universe"[7]. Simply put, here is a point where probability or chance becomes unusable. Stephen C. Meyer, P. A. Nelson, and Paul Chien wrote, "Developments in probability and complexity theory have made possible a rigorous calculation of the Universal

Probability Bound–the point at which appeals to chance in explanations become unreasonable even on a cosmic scale. In particular, the probability theorist William Dembski has recently refined the work of the earlier probabilist Emile Borel. Dembski shows that chance can be eliminated as a plausible explanation for specified systems of small probability, whenever the complexity of the system exceeds the available probabilistic (or more, precisely, specificational) resources. He then (conservatively) calculates a universal probability bound of $1/10^{150}$ corresponding to the probabilistic/specificational resources of the known universe"[8].

The main point I am trying to make is not that the probability of life occurring randomly is a very, very small number (which it is) but that we cannot even begin to put together realistic probabilities because we do not even know or have knowledge of the various steps, stages and sample sizes to realistically come up with a probability. However, we know from science that the complexity of life is great and therefore would require significant time consuming development to begin life.

Crick wrote about the messenger RNA and how many possible protein combinations there are. Using combinatorials which is mathematics involving combinations, Crick worked with an amino acid length of 200, which is less than the actual number of average proteins. Using 20 representing the number different possibilities at each place. The number of possibilities in 20x itself 200 times 20^{200}, also written as about 10^{260}, which is a number 1 with 260 xeroes [9]. Crick only considered a moderate length of polypeptide chain, not even a real life chain length, so the number of different polypeptide chains that could have been synthesized is only a very small fraction of the number of imaginable chains. Cricks calculations also only considers amino acid sequence and do not address the fact that most of the sequences would not fold up into a usable stable compact shape. Crick wrote, "What we have discovered is that even at this very basic level there are complex structures which occur in many identical copies – that is, which have organized complexity – and which cannot not have arisen by chance. Life, from this point of view is an infinitely rare event, and yet we see it teeming all around us, how can such rare things be so common"[10]. The probability and ability to prove evolutionary concepts is simply out of our reach, yet we are expected to teach our children that evolution is a fact. As this is a "fact" that we cannot fully explain, we must convolute and twist Darwin's original theory, the Big Bang model, and other related ideas just to enable us to reasonably make sense of the theory. Never mind the details – just accept what I say to be true. You don't doubt science, do you? This type of thinking reminds me of the early holiness, Pentecostal, evangelical preachers when they claimed "God is real" I cannot prove His existence but I know, that I know "God is real, real deep in my soul". I can appreciate that they knew and admitted when they did not understand things about God. These religious leader were truthful and honest.

Despite my problems with teaching evolution in school, I do not think that we should instead teach creationism or intelligent design. Let science be real science and religion be religion. I cannot provide the probability that God exists, and evolutionists cannot provide the probability that the world and life evolved.

NOTES

1. The American Cancer Society, Inc. Lifetime Risk of Developing or Dying From Cancer [Internet]. Atlanta (GA): 2014 Jan 14 [2013 Sep 5].Available from http://www.cancer.org/cancer/cancerbasics/lifetime-probability-of-developing-or-dying-from-cancer.

2. Mayo Clinic Staff. Cancer risk: What the numbers mean [Internet]. Rochester, (MN): 2014 Jan 14 [2014 Mar 1]. Available from http://www.mayoclinic.com/health/cancer/CA00053.

3. Wong, M. Creationism versus science Probability [Internet]. Toronto (CA) : 2014 Jan 14 [2014 Sep 4]. Available from http://www.creationtheory.org/Probability/Page02.xhtml

4. Answer in Genesis. A look at some figures [Internet]. Petersburg, (KY) : 2014 Jan 14 [1978 Jun 4]. Available from https://answersingenesis.org/evidence-against-evolution/probability/a-look-at-some-figures.

5. Crick, F. Life itself : its origin and nature. New York (NY): Simon and Schuster; 1981. 87 p.

6. Crick, F. Life itself : its origin and nature. New York (NY): Simon and Schuster; 1981. 87 p.

7. Morris, H. Probability and Order Versus Evolution [Internet]. Dallas (TX): 2014 Jan 14 . Available from http://www.icr.org/article/155/

8. Meyer, SC, Nelson, PA, Chien, P. The cambrian explosion: biology's big bang [Internet]. 2001; Discovery Institute. Available from http://www.discovery.org/articleFiles/PDFs/Cambrian.pdf 27 p.

9. Crick, F. Life itself : its origin and nature. New York (NY): Simon and Schuster; 1981. 51 p.

10. Crick, F. Life itself : its origin and nature. New York (NY): Simon and Schuster; 1981. 52–53 p.

Chapter 4
Cambrian Explosion

What is the Cambrian? The Cambrian represents the first geological period of the Paleozoic Era. The Oxford Dictionary defines the Cambrian period as lasting from about 570 to 510 million years ago, and it was a time of widespread seas. It is the earliest period in which fossils, notably trilobites, can be used in geological dating[1]. The Cambrian Explosion represents the period in which multicellular animal life appeared. There were major diversification of other organisms. Over a short period of time, the so-called evolutionary process accelerated and the diverse life forms started to look like those we see today. Viewing the Cambrian Explosion through evolutionary lenses, we are amazed that a wide variety of fossil forms appear almost simultaneously. There are fossil organisms that are very different in form or body plans that appear quickly. The website PBS.org gives us more information regarding the Cambrian Explosion. For about 4 billion years, life has existed on earth. During that time "evolution produced little beyond bacteria, plankton, and multi-celled algae. But beginning about 600 million years ago in the Precambrian, the fossil record speaks of more rapid change"[2]. The website adds, "Then, between about 570 and 530 million years ago, another burst of diversification occurred, with the eventual appearance of the lineages of almost all animals living today"[3].

The PBS.org website notes "The question of how so many immense changes occurred in such a short time is one that stirs scientists"[4]. Another question raised is "Why did many fundamentally different body plans evolve so early and in such profusion?" [5]. This Cambrian period shows that organisms did not need many years to develop as per Darwin's theory. The fossil record shows most of the major animal phyla are fully formed from the beginning of the geological Cambrian period. Nor is there fossil evidence that they had branched off from a so-called "common ancestor." Darwin himself does mention the Cambrian Explosion as proof *against* his theory of evolution via natural selection. Darwin was aware of this, acknowledging in *On The Origin of Species* that "I allude to the manner in which numbers of species of the same group, suddenly appear in the lowest known fossiliferous rocks"[6]. He called this a "serious" problem that "at present must remain inexplicable; and may be truly urged as a valid argument against the views here entertained"[7]. This very, very serious problem scientists have been trying to explain away (punctuated equilibrium) and to this day remains inexplicable. The views "here entertained" that Darwin is referring to are his views of evolution and natural selection as expressed in his writings. This Cambrian explosion did not happen in a twinkling of an eye; according to scientists it took over 15 million years.

Thom Holmes in his book *The Prehistoric Earth, The Early Life the Cambrian Period* wrote regarding the fossils found during the Precambrian and Cambrian that "These life-forms probably had roots as early as the Precambrian, but what makes their evolution "explosive" is the variety and number of species that evolved during the short geologic span – about 15 million years..."[8]. Pallen writing about the Cambrian Explosion asked, "How can we explain the Cambrian Explosion? The sudden appearance of so many phyla seems puzzling. However fresh new interpretations have chipped away at the taxonomic uniqueness of many Cambrian species, new fossils have closed the gap between the Cambrian and Ediacaran worlds and several molecular phylogenetic studies suggests that the diversification of animal phyla predates the Cambrian"[9]. Once again, many generalizations without the details and the so-called "chipping away" that Pallen mentions must not be too successful in terms of chiseling a clear path through the serious problems that the Cambrian Explosion poses to evolution. Darwin's writings clearly state that the process of evolution through natural selection is a slow process. Time is a very important part of his thesis. Part of the confusion is that Darwin and evolutionists use unimaginable periods of time that we cannot wrap our minds around. If given enough time, almost anything could happen especially with the generalizations presented to the general public. Darwin wrote, "That natural selection will always act with extreme slowness, I fully admit"[10]. Darwin also wrote, "...I do believe that natural selection always will act very slowly, often only at long intervals of time, and generally on only a very few of the inhabitants of the same region at the same time. I further believe, that this very slow, intermittent action of natural selection accords perfectly well with what geology tells us of the rate and manner at which the inhabitants of this world have changed"[11]. So how does this Cambrian explosion fit? *The Cambrian Explosion: Biology's Big Bang*, a paper discussing the Cambrian explosion written by Stephen C. Meyer, P. A. Nelson, and Paul Chien, notes "To say that the fauna of the Cambrian period appeared in a geologically sudden manner also implies the absence of clear transitional intermediates connecting the complex Cambrian animals with those simpler living forms found in lower strata"[12]. Another amazing part regarding the Cambrian explosion is that the fossil record shows a variety of body plans. The later geological record shows that they are stable or in stasis. Meyer, Nelson, and Chien wrote, "The major body plans that arise in the Cambrian period exhibit considerable morphological isolation from one another (or "disparity") and then subsequent "stasis." Though all Cambrian and subsequent animals fall clearly within one of a limited number of basic body plans, each of these body plans exhibit clear morphological differences (and, thus, disparity) from the others. The animal body plans (as represented in the fossil record) do not grade imperceptibly one into another, either at a given time or over the course of geological history. Instead, the body plans of the animals characterizing the separate phyla maintain their morphological isolation or disparity from all the other types of animals. They also exhibit a remarkable stability or "stasis" during their time on earth"[13].

The absence of transitional fossils in the Precambrian record has been and continues to be problematic for evolutionary theory via natural selection. Of course, scientists try to explain away the importance of such issues. Meyer, Nelson, and Chien wrote, "For many decades after the publication of *Origin*, paleontologists sympathetic to Darwin's theory sought to find the missing ancestors of the Cambrian animals. The search for the missing fossils in Precambrian formations all over the world resulted in universal disappointment. Maintaining Darwin's theory, therefore, eventually required formulating *ad hoc* hypotheses to account for the absence of ancestral and transitional forms. Various hypotheses were proposed to explain the missing ancestors, all falling under the heading of the "artifact theory." The artifact theory holds that the fossil ancestors existed, but for various reasons were not preserved in an 'imperfect and biased' fossil record. On this theory, the absence of the fossil ancestors represents 'an artifact' of incomplete sampling, and not accurate representation of the history of life. Gaps in the fossil record are apparent, not real"[14].   I find it interesting that even though the Cambrian record which we can see with our eyes reveals to us that there are no fossil ancestors we are to ignore the reality and assume it is and artifact. If we were to find on this coming week proof of fossil ancestors in relation to the Cambrian period these same scientists would forget all about artifact theory. We will read later in this book how Richard Dawkins echoes the same ideas regarding the fossil record.

The Cambrian explosion is now described as Cambrian diversification to better describe the time period. These new terms give a more accurate time description rather than implying a quick time period.

Geologic Column

I must make a brief mention of the geologic column. Thom Holm wrote, "The time it takes today for layers of the Earth to accumulate through erosion, water transport, drought, and other forces, is the same time that it took in the past"[15]. Need I ask what proof do we have that the time needed in the past is the same as it is today? We have no firsthand proof of this statement. So the first assumption that we have made is questionable. The geologic column was developed by William Smith who created the stratigraphic map. Thom Holm wrote,

"The layers were identified over time by examining their fossil contents and the positions of the layers in relation to one another"[16]. The logic used to explain the geologic column seems to be a circular reasoning and is not reliable. Even amongst scientists there were various dates given for the age of the earth. At the end of the nineteenth century, scientists agreed that the earth was many millions years old. Charles Lyell in 1860 believed the earth was 200–340 million years old. Lord Kelvin, a British physicist (1824–1907) believed the earth was 20–40 million years old. Now the earth's age is thought to be 4.5 billion years old.[17] Keeping in mind that the "geologic column" is an idea, we read "Through the widespread study of exposed rock layers all over the world, geologists have been able to visualize and record an ideal 'geologic column' representing all possible layers from the earliest rock layers to the present. Of course, there is no single place on Earth where all of these layers can be observed at one time. Earth's crust is too fragmented and twisted for that."[18]

NOTES

1. Oxford Dictionary online. Cambrian [Internet]. 2014 Jan 14: Oxford University Press; Available from Available from http://www.oxforddictionaries.com/definition/english/Cambrian.

2. WGBH Educational Foundation, Clear Blue Sky Productions, Inc. The Cambrian Explosion [Internet] : 2013 Nov 21. Available from http://www.pbs.org/wgbh/evolution/library/03/4/1_034_02.html

3. WGBH Educational Foundation, Clear Blue Sky Productions, Inc. The Cambrian Explosion [Internet] : 2013 Nov 21. Available from http://www.pbs.org /wgbh/evolution/library /03/4/ 1_034_02.html

4. WGBH Educational Foundation, Clear Blue Sky Productions, Inc. The Cambrian Explosion [Internet] : 2013 Nov 21. Available from http://www.pbs.org/wgbh/evolution/library /03/4/ 1_034_02.html.

5. WGBH Educational Foundation, Clear Blue Sky Productions, Inc. The Cambrian Explosion [Internet] : 2013 Nov 21. Available from http://www.pbs.org/wgbh/evolution/library /03/4/ 1_034_02.html

6. Darwin, C. The Origin of Species. Or the preservation of favoured races in the struggle for life [Internet]. London: 2014 Sep 26 [2013 Jan 22] Available from http://www.gutenberg.org /files/1228/1228-h/1228-h.htm

7. Darwin, C. The Origin of Species. Or the preservation of favoured races in the struggle for life [Internet]. London: 2014 Sep 26 [2013 Jan 22] Available from http://www.gutenberg.org/files/1228/1228-h/1228-h.htm

8. Holmes, T. The Prehistoric Earth : Early life The Cambrian Period. New York (NY): Chelsea House Publishers by Infobase Publishing; 2008. 166 p.

9. Pallen, MJ. The Rough Guide to Evolution. Rough Guides. London; New York: Penguin Group 2009. 163 p.

10. Darwin, C. The Origin of Species. Or the preservation of favoured races in the struggle for life [Internet]. London: 2014 Sep 26 [2013 Jan 22] Available from http://www.gutenberg. org/files/1228/1228-h/1228-h.htm

11. Darwin, C. The Origin of Species. Or the preservation of favoured races in the struggle for life [Internet]. London: 2014 Sep 26 [2013 Jan 22] Available from http://www.gutenberg. org/files/1228/1228-h/1228-h.htm

12. Meyer, SC, Ross, M, Nelson, P, Chien, P. The Cambrian Explosion: Biology's Big Bang. Discovery.org. 2001; 323-402.

13. Meyer, SC, Ross, M, Nelson, P, Chien, P. The Cambrian Explosion: Biology's Big Bang. Discovery.org. 2001; 323-402.

14. Meyer, SC, Ross, M, Nelson, P, Chien, P. The Cambrian Explosion: Biology's Big Bang. Discovery.org. 2001; 323-402.

15. Holmes, T. The Prehistoric Earth : Early life The Cambrian Period. New York (NY): Chelsea House Publishers by Infobase Publishing; 2008. 48 p.

16. Holmes, T. The Prehistoric Earth : Early life The Cambrian Period. New York (NY): Chelsea House Publishers by Infobase Publishing; 2008. 49 p.

17. Holmes, T. The Prehistoric Earth : Early life The Cambrian Period. New York (NY): Chelsea House Publishers by Infobase Publishing; 2008. 49 p.

Chapter 5
Evolutionary thought worrisome?

Dr. Cornell West uses the term nihilism to describe some of the attitudes in the black community. West describes nihilism as "lived experience of coping with a life of horrifying meaninglessness, hopelessness, and (most important) lovelessness" [1] I think that the result of accepting an evolutionary view of life will increase nihilism not only in the black community but all communities. No doubt today in America a significant amount of nihilism exists due to 911, economic hardships, natural disasters, and daily horrific news reports. An evolutionary view of life, cheapens the appreciation of human beings. In a society that already puts values on whether or not you are a man or woman, black or white, or rich or poor, evolutionary thinking would just drive us further away from equality, unity, loving each other and personal responsibility. Accepting evolutionary ideas reduces life and humanity down to chemical reactions and matter. Life and humankind is much more than a bunch of atoms and organic material.

If we are here by chance and each life is expendable and Social Darwinism – the survival of the fittest – reigns, you better not ever get sick or old. Pallen wrote "no doubt that social Darwinism was used to justify many policies that are now judged highly dubious or even morally abhorrent:"[2]. Ideas like eugenics condoned the sterilization of the "less fit" members of society. We must not forget the Nazis in the 1930–40 sterilized those that were unfit but also locked up the disabled. Finally Hitler destroyed and killed many during the

Holocaust. The bottom line here is, like the late Bishop Ithiel Conrad Clemmons once said during one of his sermons, "Life looks one way when you're young, but when you get older it looks another way"[3] . What he was saying is that when your health and economic conditions are fine, you will see life one way. When we get older and have experienced the vicissitudes of life such as disappointments, failing health, loss of a job or people we love; sometimes our view of life begins to change. Concepts like survival of the fittest are attractive when things are going well. However, when the first sign of health problems due to a car accident, depression, anxiety, or stress occurs or when natural selection produces a genetic mutation in the form of cancer, then what? We realize that we can't pull ourselves up using our own strength, that we need a cane or a helping hand. No doubt the strong and healthy group that we once were a part of will cease to socialize with us. Surely intellectual reasoning alone will not help when life throws us physical, mental and economic curves. We will need help and each other; life is designed this way.

Jonathan Clements in *Darwin's Notebook The Life, Times, and Discoveries of Charles Robert Darwin*, under Social Darwinism, wrote, "A number of European thinkers particularly the ardent Zoologist Ernst Haeckel, began to debate the effect of Darwin's theories on human society. If fortune favored the fittest, what was 'fittest' for society? The debates soon led to ward notions of racial superiority. Where Darwin had emphasized the similarities between modern human races, his cousin Francis Galton began to dwell on the differences"[4]. Francis Galton wanted to send the Chinese to settle in Africa. He believed that the Chinese would continue to exist but would eventually take the place of the inferior Africans. Galton's letter to the Times, on June 5, 1873, stated, "I should expect the large part of the African seaboard, now sparsely occupied by lazy, palavering savages... might in a few years be tenanted by industrious, order loving Chinese, living either as a semi–detached dependency of China, or else in perfect freedom under their own law"[5].

Most people are not aware that the Bronx Zoo once exhibited an African man named Ota Benga along with an orangutan in 1906. Mr. Ota Benga, a Congolese pygmy, was 4-feet-11-inches tall. His family wife and children were slaughtered by the Force Publique, an armed force. He was forced into slavery and was discovered by an anthropologist Samuel Philips Verner who wanted to use him in an anthropology exhibit. Mr. Ota Benga went to New York with Mr. Verner in August 1906 as part of his collectibles. Eventually, Mr. Ota Benga spent his day in the zoo along with the other animals, particularly monkeys, and an orangutan named Dohong. The zoo personnel wanted Mr. Benga to spend time with the animals, he had his hammock in the caged area. The exhibit was titled "Bushman Shares a Cage With Bronx Park Apes." Visitors came to see Mr. Ota Benga in the cage along with the orangutans. The New York Times story mentions a group of black ministers and their reactions. "To the black ministers and their allies, the message of the exhibit was clear: The African was meant to be seen as falling somewhere on the evolutionary scale between the apes with which he was housed and the people in the overwhelmingly white crowds who found him so entertaining"[6]. Also, a Mr. Gordon who was from the Howard Colored Orphan Asylum, said that the exhibit was not only racist but seeks to show Darwin's theory of evolution. Evolutionary thought has indirectly produced many ills which still affect us today. Ernest Haeckel supported a theory called recapitulation that supported the belief that "humankind itself had lower and higher orders and that Haeckel's own race, the white European, occupied a superior branch on the tree of life to say, than that of the black African. Today his theories are widely rejected"[7]. Haeckel was a skilled artist, but he deceitfully drew illustrations of embryos that showed an evolutionary progression from a common ancestor. However, his theory was used to support the idea that biology determines social position. Jonathan Clements wrote, "If white culture and white civilization were considered 'superior,' then it followed that no social, financial or educational remedy could rescue 'lower races' from their biological inferiority"[8]. Haeckel lived from 1834–1919, which was during the time of slavery and its aftermath. Scientists were discussing evolution, and pseudo–scientists were trying to justify human slavery. Haeckel either was ignorant or ignoring the history of the black African because they were among the leaders of the world in the 15th century with civilizations equal to and in some ways greater than those in Europe at that time. Their civilization was a truly black African civilization, wholly non-European, and not the product of Islam or some other outside culture. W.E.B. Dubois in his book *The World and Africa*, quoted Frobenius, "The revelation of fifteenth and seventeenth century navigator furnish us with certain proof that Negro Africa, which extended south of the Sahara desert zone, was still in full bloom, in the full brilliance of harmonious and well–formed civilizations. In the last century the superstition ruled that all high culture of Africa came from Islam. Since then we have learned much, and we know today that the beautiful turbans and clothes of the Sudanese folk were already used in Africa before Muhammed was even born or before Ethiopian culture reached inner Africa. Since then we have learned that the peculiar organization for the Sudanese states existed long before Islam and that all of the art of building and education, of city organization and handwork in Negro Africa, were thousands of years older than those of Middle Europe"[9].

Eugenics is also an offshoot of Darwinism. The Oxford Dictionary defines Eugenics as "the science of improving a population by controlled breeding to increase the occurrence of desirable heritable characteristics." Clements wrote, "The eugenics movement found support all over the world. In the United States, Alexander Graham Bell published the results of his own investigations into the congenital deafness in Martha's Vineyard, concluding that two deaf parents were more likely to have deaf children. He subsequently recommended that partners with matching disabilities should be discouraged from marrying. The Eugenics Record Office (ERO) began in 1910 watching families' pedigrees, which led to state programs that stopped the immigration of so called "undesirables," mixing of races, and individuals that would have so called "unfit children"[10]. America led out with eugenics that was focused mostly on "Nordic" races, which the Oxford Dictionary defines as "relating to or denoting a physical type of northern European peoples characterized by tall stature, a bony frame, light colouring, and a dolichocephalic head"[11]. America's eugenics considered those of Eastern European (the portion of the European landmass that lies east of Germany and the Alps and west of the Ural Mountains, including the former Eastern bloc countries of Poland, the Czech Republic and Slovakia, Hungary, Romania, and Bulgaria, as well as the Baltic republics of Estonia, Latvia, and Lithuania, and the former Soviet republics of Belarus and Ukraine, along with Russia west of the Urals), black African, and Asian heritages as undesirable[12]. In Australia, children that were half aborigine were taken from their parents so that they could be raised as white persons[13].

Then came one of the greatest evil offshoots of Darwinism – fascism and Adolf Hitler in Germany. Hitler believed that the Germans were a "master race" that needed to be preserved and protected from the negative effects of miscegenation. Hitler tried to promote positive eugenics, which was the continued growth of the German children with their positive traits. Hitler tried to eliminate the "undesirables" through negative eugenics death, abortion and destruction[14]. I do not know where Hitler was going to get this pure white race because we are all mixed and have ancestors that at one time have had brown or black skin, which he considered unacceptable.

Religion and the belief in God have been shown to have benefits in everyday life. In fact, a study led by researchers at the Dana–Farber Cancer Institute and Harvard Medical School showed that spirituality played an important role in coping with cancer. These patients revealed that they were mostly left without religious support: "...multi–institutional investigation of advanced cancer patients and their main caregivers. Of 230 patients surveyed, the vast majority – 88 percent – considered religion to be at least somewhat important. But nearly half said their spiritual needs were largely or entirely unmet by a religious community, and 72 percent felt those needs were similarly unaddressed by the medical system"[15]. This is sad because findings revealed that religious support was almost nonexistent. "The findings also indicated that greater spiritual support from religious organizations and medical service providers was strongly linked to better quality of life for patients, even after other factors were taken into account. Intriguingly, patients who considered themselves religious were more likely to want all possible measures taken to extend their lives"[16]. This shows that the patients were willing to try and therefore live longer by testing other treatment options. Tracy Balboni, MD and senior resident at Harvard Radiation Oncology Program, said, "Our findings suggest that such support can help improve patients' quality of life at the end of life"[17]. These highly religious patients no doubt helped themselves, doctors and scientists by their willingness to accept all the help they could to get well. "The finding that highly religious patients were the most likely to desire life–extending measures came as something of a surprise, said Balboni. Such individuals might be expected to submit to the natural unfolding of a divine plan, rather than want heroic measures. But, the authors suggested, 'Religious individuals may feel that because their illness is in divine hands, there is always hope for a miraculous intervention. Religious individuals also may place a value on life that supersedes potential harms of aggressive attempts to sustain life' "[18]. Belief in God does not mean that science is irrelevant and doctors are not necessary.

In another study by the Rush University Medical Center showed that patients who were diagnosed with clinical depression and believed in a concerned God had improved response to their medical treatment. The paper, published in the Journal of Clinical Psychology, surveyed 136 patients with major depression or bipolar depression in Chicago inpatient and outpatient facilities. "...the study found that those with strong beliefs in a personal and concerned God were more likely to experience an improvement. Specifically, participants who scored in the top third of the Religious Well-Being Scale were 75 percent more likely to get better with medical treatment for clinical depression. The interesting thing was that the study went further to see whether or not it was due to hopefulness"[19]. Patricia Murphy, a chaplain and assistant professor of religion, health and human values at Rush University, wrote, "In our study, the positive response to medication had little to do with the feeling of hope that typically accompanies spiritual belief. It was tied specifically to the belief that a Supreme Being cared"[20]. This is a scientific study that shows the mere belief in a caring God, not wishful, hopeful thinking was the cause of improvement. I am not suggesting that doctor's initiate or try to convert patients to any religion but merely support the patients overall. I remember watching a movie a few years ago about the South after the end of the Civil War. Blacks were fighting for their rights using the law and the church. There was an upcoming court trial in the South which the blacks hoped that the jurors would "do justice" and convict the racist persons responsible for the crime. There was a scene where another black person from the town went to the pastor of the church where most of the blacks attended and asked him why he encouraged and built the faith of these poor black people when you know that the all white jurors are not going to be fair? The people are going to be let down again, after all. So why do it? I recall the pastor looking very concerned and with a painful look on his face explaining that if he did not continue to build the people's faith even though he knew very well that justice would not prevail, the people would lose all hope. What I got out of the pastor's statement was that when life is closing in all around you and everything seems hopeless, it is your faith in something other than yourself (God) that will keep you going and life's problems will not break your spirit. When the spirit is broken you are finished. The Bible states that no man can endure a broken spirit: "The spirit of a man will sustain his infirmity; but a wounded spirit who can bear?" (Proverbs 18:14)

In the book *Tortured for Christ*, Rev. Richard Wurmbrand – a Jew and former atheist who converted to Christianity – wrote about his time in Romania where starting on August 23, 1944, one million Russian troops entered his country. After disarming the army and public, the communists came to power. The communists initiated a congress at which all Christian groups came together. There were over four thousand pastors and ministers in attendance.[21] At this meeting, Stalin was elected president. Soon after these religious leaders were under the control and influence of communism, some wore robes with the sickle and crescent sign. The church had to mostly go underground to prevent beatings and imprisonment. Torture included brainwashing in which the prisoners spent sometimes seventeen hours a day for weeks and months hearing, "Communism is good, Christianity is stupid, Give up!"[22]. The communists often said "You know, I am God. I have power of life and death over you. The one who is in heaven cannot decide to keep you in life. Everything depends upon me. If I wish, you live. If I wish, you are killed. I am God"[23] . Even though the communists had control over the people's property, lives, wealth and food, they knew that they would not be in total control until they broke their spirit by eliminating all faith in God. They understood the power of faith.

The reason I mention this is to point out that dictators and evil rulers know that they must eliminate all ideas of God if they are to be successful in controlling the masses. Perhaps the communists recognized the power and strength that the people had when they held onto their faith in God. So they tried to brainwash, beat, and maim, to get the people to willingly renounce their faith.

There are also psychological implications of evolutionary thinking, especially as it relates to genes and our obsession with health and death. When we have genetic testing done at the doctor's office or lab, we provide a sample of blood or saliva. However, when we get the results back and we have proof of a genetic mutation known to be partly responsible for cancer or another sickness, sometimes our lives are in upheaval. These tests not only affect the person involved but also can affect our relationships. How do we handle the news when we find out that we have a mutated gene that is responsible for breast, colon or prostate cancer? Sometimes we allow negative information to strip us of life today by worrying about something that probably might never occur.

I am amazed at the way companies use "science" to their profit. The company www.23andme.com provides health-related genetic testing for various diseases and ancestral origins for about $99. People would send in saliva samples and the company would test the sample and tell people whether or not they are predisposed to certain diseases. However, recently the Food and Drug Administration forced them to discontinue their action while they are being reviewed. On their website, you will see this statement regarding their health-related genetic testing: "Customers who purchased or have purchased 23andMe's Personal Genetics Service (PGS) on or after November 22, 2013, the date of the Warning Letter from the FDA, will receive ancestry information, as well as their raw genetic data without interpretation. These new customers may receive additional health related results in the future, dependent kits on or after November 22, 2013 will be eligible for a refund"[24]. This was done because the company failed to prove the validity of its product. The FDA said that it did not have any assurance that the company analyzed or clinically validated the personal genome service for its customers. It was pseudo-science used to make money from people by playing to their fears and paranoia. The interesting thing is the people that purchased these test were not only the people with or without a high school diploma but also a board-certified dentist.

Evolutionists in the past have used and presently use so called vestigial organs to support their theory. The Oxford Dictionary defines vestigial as "*Biology* (of an organ or part of the body) degenerate, rudimentary, or atrophied, having become functionless in the course of evolution"[25]. *The Evolution of Life* defines a vestigial organ as one "which has lost its function in the course of evolution, and is usually much reduced in size." Today we see the revised definition that is so vague. The fact is that the organs once thought of as proof of evolution – like the appendix, tonsils, pineal gland, and thymus

– actually are all are very important to the function of the organism[26]. These organs were taken out of the human body because they were assumed to be useless and did not perform a function. These assumptions were based partly upon evolutionary ideas that were incorrect. How many tonsillectomies were done because they were enlarged or infected? Doctors swayed by evolutionary ideas thought that if we take them out through surgery we can improve health. After all they are no longer useful because they are vestigial. We found out that the reverse had actually happened. There were more upper respiratory infections after tonsil surgery. The so called vestigial organs like the appendix which once were numbered over 150, have been greatly reduced. This is because modern science has found that these organs are being used to support the life of the organism. A study reported that the appendix was partly responsible for the development of ulcerative colitis in young patients with ulcerative appendicitis. The study also suggested that further studies should be done to further clarify the role in the pathogenesis of the ulcerative colitis[27]. According to Loren G. Martin, a professor at Oklahoma State University, "For years, the appendix was credited with very little physiological function. We now know, however, that the appendix serves an important role in the fetus and in young adults. Endocrine cells appear in the appendix of the human fetus at around the 11th week of development…"[28]. Doctors thought that "silent" mutations were unimportant to the health of an organism. Medical professionals have even termed the phrase "silent" mutations, which are mutations that were believed to not have a significant effect on the human body[29]. The changes or mutations would not cause alterations to the composition of the proteins which are encoded by genes. We are now finding out that we were wrong these so called "silent" mutations do indeed affect the health and operation of the organism.

In the movie "American Addict," a documentary on the United States and the people's appetite for prescription drugs by Gregory Smith (director, Sasha Knezev), Peter Breggin, M.D. and author of the book *Toxic Psychiatry*, said that the most successful public relations campaign was done by Eli Lilly when it said that depression was caused by a biochemical imbalance that Prozac could fix. The speaker mentioned that in the past if a young sophomore in college was struggling with angst, depression or other life problems, the student would read a deep book, seek counseling, become religious, study philosophy, or Freud but ultimately would view himself as struggling with life issues[30]. Today, however, the college student would be told that depression or every problem can be reduced to biochemical imbalance. The focus today is that all the anguish people feel is due to biochemical imbalance and with this conclusion all of the learning of Western civilization is thrown out[31]. "This is an outrageous societal phenomenon and it indicates a culture losing touch with empathy for itself as individuals and empathy for other human beings, a culture that has just gotten caught up in the end product and efficiency at work; and not at all thinking about what is life about, who am I, who are my children, are my children gifts from God; are they to be treasured; do they require that I change my life, bring them up better or are they objects to be tinkered with and given drugs."[32] After watching this documentary, I began to think about the way evolutionary thinking produces similar results. Science and medication in tandem convince us that we are in free fall and no one is in control of his or her own life. This feeling of hopelessness is crippling and a defeatist attitude.

NOTES

1. West, C. Race Matters. Boston (MA): Beacon Press; 1993. 14 p.

2. Pallen, MJ. The rough guide to evolution. Rough Guides. London; New York: Penguin Group; 2009. 266 p.

3. Clemmons, IC. Sermon presented at The Historic First Church of God in Christ. Brooklyn, NY.

4. Clements, J. Darwin's notebook the life, times, and discoveries of Charles Robert Darwin. China; Quid publishing; 2009. 146 p.

5. Clements, J. Darwin's notebook the life, times, and discoveries of Charles Robert Darwin. China; Quid publishing; 2009. 146 p.

6. Keller, M. The scandal at the zoo. [Internet]. New York (NY): The New York Times. 2006 Aug 6 [cited 2014 Sep 26]. New York Region. Available from http://www.nytimes.com /2006/08/ 06/nyregion/thecity /06zoo.html?pagewanted= all&_r=1& .

7. Clements, J. Darwin's notebook the life, times, and discoveries of Charles Robert Darwin. China; Quid publishing; 2009. 146 p.

8. Clements, J. Darwin's notebook the life, times, and discoveries of Charles Robert Darwin. China; Quid publishing; 2009. 146 p.

9. Dubois, WEB. The world and Africa: An inquiry into the part which African has played in world history. New York; International Publishers; 1965. 210-202 p.

10. Clements, J. Darwin's notebook the life, times, and discoveries of Charles Robert Darwin. China; Quid publishing; 2009. 148 p.

11. Oxford Dictionary online. Nordic [Internet]. 2014 Jan 14: Oxford University Press; Available from Available from http://www.oxforddictionaries.com/definition/english/Nordic.

12. Clements, J. Darwin's notebook the life, times, and discoveries of Charles Robert Darwin. China; Quid publishing; 2009. 148 p.

13. Clements, J. Darwin's notebook the life, times, and discoveries of Charles Robert Darwin. China; Quid publishing; 2009. 149 p.

14. Clements, J. Darwin's notebook the life, times, and discoveries of Charles Robert Darwin. China; Quid publishing; 2009. 149 p.

15. Holly, P., Block, S., Lathan, C., Peteet, J., Vanderwerker, L., Paulk, E. Study recommends greater attention to spiritual needs of people with advanced cancer [Internet]. Boston, (MA): Dana–Farber Cancer Institute; 2013 Aug 8 [Cited 2007 Feb 7]. Available from http://www.dana–farber.org/ Newsroom/News–Releases/Study–recommends–greater– attention–to–spiritual–needs–of–people–with–advanced– cancer.aspx.

16. Holly, P., Block, S., Lathan, C., Peteet, J., Vanderwerker, L., Paulk, E. Study recommends greater attention to spiritual needs of people with advanced cancer [Internet]. Boston, (MA): Dana–Farber Cancer Institute; 2013 Aug 8 [Cited 2007 Feb 7]. Available from http://www.dana–farber.org/ Newsroom/News–Releases/Study–recommends–greater–

attention–to–spiritual–needs–of–people–with–advanced–
cancer.aspx.

17. Holly, P., Block, S., Lathan, C., Peteet, J., Vanderwerker, L.,
Paulk, E. Study recommends greater attention to spiritual
needs of people with advanced cancer [Internet]. Boston,
(MA): Dana–Farber Cancer Institute; 2013 Aug 8 [Cited 2007
Feb 7]. Available from http://www.dana-farber.org/
Newsroom/News-Releases/Study-recommends-greater-
attention-to-spiritual-needs-of-people-with-advanced-
cancer.aspx.

18. Holly, P., Block, S., Lathan, C., Peteet, J., Vanderwerker, L.,
Paulk, E. Study recommends greater attention to spiritual
needs of people with advanced cancer [Internet]. Boston,
(MA): Dana–Farber Cancer Institute; 2013 Aug 8 [Cited 2007
Feb 7]. Available from http://www.dana-farber.org/
Newsroom/News-Releases/Study-recommends-greater-
attention-to-spiritual-needs-of-people-with-advanced-
cancer.aspx.

19. Rush University Medical Center. Belief in a Caring God
Improves Response to Medical Treatment for Depression
[Internet]. Chicago (IL): Rush University Medical Center ; 2013
Aug 11 [2010 Feb 23]. Available from http://www.rush.edu
/webapps /MEDREL/servlet/News Release ?id=1353.

20. Rush University Medical Center. Belief in a Caring God
Improves Response to Medical Treatment for Depression
[Internet]. Chicago (IL): Rush University Medical Center ; 2013
Aug 11 [2010 Feb 23]. Available from http://www.rush.edu/
webapps/ MEDREL/servlet/News Release ?id=1353.

21. Wurmbrand, R. Tortured for Christ. Bartlesville, (OK):
Living Sacrifice Book Company; 2013. 14 p.

22. Wurmbrand, R. Tortured for Christ. Bartlesville, (OK): Living Sacrifice Book Company; 2013. 41 p.

23. Wurmbrand, R. Tortured for Christ. Bartlesville, (OK): Living Sacrifice Book Company; 2013. 45 p.

24. 23andMe. Status of our health-related genetic reports [Internet]. Mountain View (CA): 2014 Jan 14. Available from https://www .23andme.com/health/

25. Oxford Dictionary online. Vestigial [Internet]. 2014 Jan 22: Oxford University Press; Available from Available from http://www.oxforddictionaries.com/definition/english/Oxford/vestigial

26. Gamlin, L, Vines, G. The Evolution of Life. New York (NY): Oxford University Press, Oxford University Press; 1986.

27.   Matsushita, M, Uchida, K, Okazaki, K. Role of the appendix in the pathogenesis of ulcerative colitis. Inflammopharmacology. 2007 Aug ;15 (4): 154-157. doi 10.1007/s10787-007-1563-7

28. Scientific American. What is the function of the human appendix? Did it once have a purpose that has since been lost? [Internet]. U.S. Canada: 2014 Sep 24 [1999 Oct 21]. Available from http://www.scientificamerican.com/article/what-is-the-function-of-t/

29. Chamary JV, Hurst LD. The price of silent mutations. Scientific American. 2009; 300(6): 46-53.

30. Knezev, S, Smith, G. American Addict (Netflix Movie). Peter Breggin, M.D. Author of the book Toxic psychiatry.

31.   Knezev, S, Smith, G. American Addict (Netflix Movie). Peter Breggin, M.D. Author of the book Toxic psychiatry.

Chapter 6
Life itself & Cells

Francis Crick, a molecular biologist, biophysicist and neuroscientist, co-discovered the structure of DNA molecules in 1952 and was jointly awarded the Nobel Prize for Physiology or Medicine. In his book *Life Itself,* Crick expanded on the idea that Svante August Arrhenius, a Swedish chemist and physicist, had about the origin of life on earth. Arrhenuus, who received a Nobel Prize for Chemistry in 1903, proposed that life did not start on earth but was seeded by a microorganism wafted in from space.[1] He called this concept *panspermia,* meaning "seeds everywhere." Crick in *Life Itself* suggests along with Leslie Orgel their slightly varied idea of directed panspermia. Which is that an organism travelled at the front of an unmanned spaceship which was sent by a higher being/creature. This higher civilization developed somewhere other than earth billions of years ago. Life started here on earth when the organisms were dropped into the primitive ocean and multiplied. They published their idea of panspermia in *Icarus.* Simply put, Crick suggested that life was started here on earth because an alien/intelligent being – not chance – sent an unmanned speedy spacecraft with an organism attached to the front of spaceship. This spaceship flew very, very fast and was able to slow down to a speed that allowed the organisms attached to be placed into water here on earth. These organisms multiplied, and through evolution we are here today.

Crick mentions the Italian Physicist Enrico Fermi, who asked the question during his speech "So where are they?"[2] He was referring to the intelligent being/force that is responsible for life here on earth, as suggested by other scientists like Arrhenius. Crick then gives us a glimpse of some of the things that Fermi mentioned during his speech about the beginning of life by listing a series of steps: *a.* Our galaxy has many stars, there are 10^{11} galaxies or more. Many of these stars *may* have planets circling them, and a small number of them would have liquid water and a gaseous atmosphere, carbon, nitrogen, oxygen hydrogen. *b.* The sun's rays would cause the surface of planet to synthesize organic compounds, this would result in the ocean turning into a soupy mixture. Chemicals begin to join and combine and would interact in complicated ways that would produce a self-producing system or a primitive form of life. *c.* These primitive life forms would increase and then evolve by natural selection to a complex organism with the ability to think. *d.* Civilized, scientific and technological concepts would follow, leading to a search to discover new worlds. These highly advanced beings eventually would discover the earth's many advantages, such as temperature, water and chemicals. After this short list explaining the general outline with the addition of panspermia, Fermi then asked the question "Where are they?" Why have we not yet seen them or know them?

Crick, writing about Fermi's general outline above, brings out the inability of scientists to come up even with probabilities of Fermi's steps actually occurring. There are so many unknown variables that we cannot begin to come up with (p) probability. Crick wrote, "Most people would accept the general trend of Fermi's argument. The difficulties arise when one tries to estimate the probability of each step, to put in numbers. There is no really hard evidence that other stars have planets, although it certainly seems likely that they do. If planets exist, at least a few will probably have a favorable environment for the production of a good soup – a mixture of simple organic compounds in water."[3]

Finding these facts printed in your newspaper or on a website that gives the reader a balanced understanding of evolutionary ideas would be tedious. In light of what we just do not know, on December 10, 2013, Ron Cowen wrote a story in Nature Magazine titled *Simulations back up theory that Universe is a hologram* and reported that physicists presented evidence that our universe may be a projection. Cowen wrote, "A team of physicists has provided some of the clearest evidence yet that our Universe could be just one big projection."[4] The details of the story I will not go into here, but we see that our understanding is very limited and uncertain. You have one group of scientists saying that what we see in our large telescopes is just a big visual illusion similar to a very large 3D format screen. We are not even sure what we see, but we are *sure* (mostly) that the universe with its stars and planets are real.

The general public is fed these ideas as if there are no serious unanswered questions that could strengthen or weaken the theory. Yet Crick wrote about the step that evolutionists gloss over like life created from non-life. Crick says of the chemical activities in Fermi's steps (b, c above), "It is the next step which at present so mysterious: the formation from the soup of a primitive, chemical self-reproducing system."[5] Crick mentions that even if these events did happen, we do not know how likely it is for the long process of evolution to reach a higher civilization or whether these so-called higher beings or creatures (aliens) would explore the universe or how far they would actually be able to travel. Crick states that Fermi's events may be happening, but the steps would be rare and very, very slow. But enough about these fantastic ideas, I will stick with the SciFi channel, where at least it seems more realistic.

Crick, writing about the probability of abiogenesis, he said that the exact figures do not matter. Crick wrote "that we have no idea what value we should take for p (probability), except that it should be 'small.' For this reason, it is impossible for us to decide whether the origin of life here was a very rare event or one almost certain to have occurred"[6] Crick is probably the most upfront writers of evolutionary concepts I have read. Crick wrote, "An honest man, armed with all the knowledge available to us now, could only state that in some sense, the origin of life appears at the moment to be almost a miracle, so many are the conditions which would have had to have been satisfied to get it going."[7]

Crick also wrote about defining what is considered *life* or *living*, defining what is alive – plants, bacteria, viruses. Viruses, however, are on a borderline of living versus nonliving. When we look at the very small, like viruses and bacteria, we see very complex operation. Crick wrote, "When we do this, we cannot help being struck by the very high degree of organized complexity we find at every level, and especially at the molecular level. Since we have every reason to believe that structures easily visible to the naked eye as well as those seen only with a microscope are all built up from the intricate interactions of their molecular components."[8] Evolutionary thought requires that we evolved from the simple to the advanced in stages or steps over very, very long periods of time. Darwin wrote in *On the Origin of Species* 1st Edition, "If it could be demonstrated that any complex organ existed, which could not possibly have been formed by numerous, successive, slight modifications, my theory would absolutely break down. But I can find out no such case. No doubt many organs exist of which we do not know the transitional grades, more especially if we look to much-isolated species, round which, according to my theory, there has been much extinction." [9] When looking at the cellular level of life, we see extremely complex factories which are irreducible, and their evolutionary steps cannot be completely understood today.

Crick mentions macromolecules, specifically the protein family. The most simple proteins can have well over 2,000 atoms. Each atom is in a particular location and becomes moved or changed when they are exposed to heat while in a solution. The increased temperature will loosen the weak bond that holds the chain in the correct fold. Crick mentions that nature synthesizes the polypeptide chain in a one dimensional form and then it folds up into the three dimensional form. This is very complex and I bet that no detailed evolutionary explanation exists as to how and why this chemically related folding up process evolved. Crick wrote, "To produce this miracle of molecular construction all the cell needs to do is to string together the amino acids (which make up the polypeptide chains) in the correct order. This is a complicated biological process, a molecular assembly line, using instruction in the form of a nucleic acid tape (so called messenger RNA)." [10]

As I read books regarding evolution, the authors often present information as if there are no questions surrounding evolution. Crick expressing his thoughts about non–life begetting life wrote, "The plain fact is that the time available was too long, the many microenvironments on the earth's surface too diverse, that various chemical possibilities too numerous and our knowledge and imagination too feeble to allow us to be able to unravel exactly how it might or might not have happened such a long time ago, especially since we have no experimental evidence from that era to check our ideas against."[11] Crick confirms that we do not have any evidence concerning the environment, earth's surface, and time for development that would enable us to come to a firm conclusion regarding life origins. Richard Dawkins, an evolutionary biologist and atheist, tells us that evolution is a fact, and he tries to convince us that we do not have a right to even expect evidence from the earth. Any evidence that we do have should consider bonus material. Crick tells us that evidence of the earth's environment and experimental proof would be necessary to demonstrate our theories. However, Richard Dawkins tells us that we should not expect to find fossil evidence.

Crick wrote about extrasolar planets and the lack of evidence or proof of their existence. Scientists use the Doppler Effect to determine the rate of spin of stars. The slower spinning stars have planetary systems.[12] Crick said of the Doppler Effect, "Unfortunately, this is really the only evidence we have for the existence of planets. One is always more comfortable in science if two or more distinct lines of reasoning lead to the same conclusion. Here we have only one. Experience has shown that such a deduction can only be accepted with reserve. Having said that, one must concede that the direct evidence for stellar rotations is really very convincing, the deduction about the existence of planets circling around slow-spinning stars fairly plausible and not incompatible with our broad theories as to how stars and planets may have arisen." [13] Clearly we see that there is evidence, although it is very limited and therefore should be accepted with caution. This is not to say we should throw away what we do know but that we must consider it appropriately.

Returning to panspermia, Crick alerted us in his book *Life Itself* that he would be presenting ideas that were not scientifically verified: "From this point on we must leave behind quantitative considerations however approximate, and allow our imagination a somewhat freer hand. We shall postulate that on some distant planet, some four billion or so years ago, there had evolved a form of higher creature who like ourselves, had discovered science and technology." [14] Crick said that these creatures would have used an unmanned spaceship to probe the earth. In the front of the unmanned spaceship would be an organism, probably blue-green algae. The algae is supposed because of the early fossil record of evolutionary trees, as the earlier fossils resemble the blue-green algae. [15] Crick points out that even the evidence that supports this idea of using blue green algae is weak because "We do not have available a whole series of sedimentary rocks dating back from 3.6 billion to 4.6 billion years before the present era, or thereabouts. Thus it is not surprising that we lack the evidence for earlier forms." [16] The alien spaceship could go very, very fast and would be able to land without damaging the organism in the front possibly frozen state–organisms. The spacecraft would have to work for tens of thousands of years. This higher creature/alien directed the spaceship to provide interplanetary fertilization (*Directed Panspermia*). I am amazed at the way scientist can suggest and even publish ideas that the average person would be considered crazy for suggesting. It seems as long as we stay within the bounds of evolutionary thought, we can suggest almost anything as long as a few scientific principles support at least some part of our idea. In *The Rough Guide to Evolution,* Pallen wrote "The fledgling field of astrobiology brings an inter–disciplinary slant to the potential for life in the universe. There is as yet no direct evidence for the existence of life outside of Earth."[17]

NASA's two robot geologists Mars Exploration Rovers, launched for Mars on June 10 and July 7 in 2003, were given the mission to search for water on Mars. They landed in January 2004. On December 15, 2013, NASA's rover Curiosity provided evidence from chemical analysis of material drilled by Curiosity. Mars the Red Planet could have supported life billions of years ago. There was mud found on Mars, which indicates that a lake might have existed in the area. However, there was also high levels of radiation found on Mars. Jennifer Eigenbrode a biochemist and geologist at NASA said, "We don't know if life on Mars could have ever adapted to the high levels of radiation the surface is currently experiencing."[18] So far, what we have is similar to the Urey–Miler experiment assumption of water present, a few chemicals that are helpful to support life. But – and a big but – radiation is present and in dangerous amounts. Radiation is harmful to human and other living things. It is more than likely that organisms would not be able to exist in such a harsh environment. The National Academies National Research Council Division on Earth and Life Studies Board on Radiation Effects Research reported on

June 28, 2005, about the effect of low dose exposure to radiation. The report concluded that there is no safe level or threshold of ionizing radiation exposure. Exposure to background radiation can cause cancer. Additional exposure can also cause additional risks. More than likely primitive life on Mars does not stand a chance.[19]

I was amused by the question Seth Shostak asked, "Why Doesn't NASA Just Look for Life?" Shostak, the senior astronomer at the SETI Institute wrote, "The 'water found on Mars' story is as perennial as Christmas. NASA doesn't need to tell us that again. So why not cut out the seemingly endless stream of robotic middlemen, and just send hardware that will search for life, big or small?" [20] I think I know the answer – they know that there is no possible life based upon the environment. More than likely, there will be no proof or evidence that would be detectable.

Crick himself mentions the improbability and increased speculation regarding life origins. "I write a paper on the origins of life I swear I will never write another one, because there is too much speculation running after too few facts, though I must confess that in spite of this, the subject is so fascinating that I never seem to stick to my resolve."[21] At least Crick lets the reader know when scientific concepts are on shaky ground so the reader will not be easily misled in thinking that all scientific ideas are solidly verifiable.

Simon Millton's book *Conflict in the Cosmos Fred Hoyle's Life in Science.* Sir Fred Hoyle was an English astronomer known for stellar nucleosynthesis. He also wrote science fiction along with his son Geoffrey Hoyle. He was born on June 24, 1915, and he died after medical complications in 2001. Hoyle often disagrees with cosmological theories of life origins. He was an atheist; however, his work on the carbon nucleus caused him to ask certain questions that seemed to contradict his atheism. Fred Hoyle wrote in a 1981 issue of Cal Tech alumni magazine, "Would you not say to yourself, some super calculating intellect must have designed the properties of the carbon atom, otherwise the chance of my finding such an atom through the blind forces of nature would be utterly miniscule. Of course you would... A common sense interpretation of the facts suggest that a super intellect has monkeyed with physics, as well as with chemistry and biology, and that there are no blind forces worth speaking about in nature. The numbers one calculates from the facts seem to me so overwhelming as to put this conclusion almost beyond question."[22] Hoyle's comments remind me of the biblical scripture in Romans chapter 1 "[19]Because that which may be known of God is manifest in them; for God hath shewed it unto them. [20]For the invisible things of him from the creation of the world are clearly seen, being understood by the things that are made, even his eternal power and Godhead; so that they are without excuse: [21]Because that, when they knew God, they glorified him not as God, neither were thankful; but became vain in their imaginations, and their foolish heart was darkened."[23] I understood these verses to mean that men and women of learning have known for a long time what Hoyle wrote above, that there is another force to be reckoned with. How we deal with this information is up to each individual. The bible mentions that we understand things in parts and do not have the complete picture. There are things that we simply do not and cannot understand at this time. This is not to say that we should fold our hands and say, "Oh well, that's the way it is," but we should seek for the understanding. In fact, the Bible clearly states that if one lacks understanding, ask for it, don't just sit back and do nothing. Faith is belief in spite of what we see and feel as it's having faith in what the Bible states is true. The Bible states in Hebrews 11:1 "Now faith is the substance of things hoped for, the evidence of things not seen." Also in 2 Corinthians 4:17–18 we read, "For our light affliction, which is but for a moment, is working for us a far more exceeding and eternal weight of glory, while we do not look at the things which are seen, but at the things which are not seen. For the things which are seen are temporary, but the things which are not seen are eternal." [24] Faith in God is not an excuse to evade thinking and evaluation. In fact, faith in God prompts deep thinking, meditation and repeated evaluation of God's message in the Bible. There is a deeper focus on things that are not self-centered. Faith requires us to work toward improvement despite the odds and the seemingly endless battles. We pray, seek and believe God for change while we continue to work toward that change. It is the very faith that we live by. I think today that the world views faith through distorted lenses caused by non-biblical thinking and poor Christian examples. I do not expect most people to understand because unless one has experienced real life-changing faith, one cannot possibly understand faith. Faith in God prompts one to question even more and drives one to higher levels of service, commitment and personal growth.

Earth's galaxies

In the book *The Cell, Evolution of the First Organism*, Joseph Panno, PhD., wrote that "Life began so long ago that many people believe it is impossible to reconstruct the events that led to the appearance of the first cell."[25] No doubt many people believe this and with good reason. The length of time that is considered necessary is simply unimaginable. We can barely confirm what happened 100 years ago even with our technology. Panno then starts to explain the general outline of the Big Bang and explains that about 15 billion years ago everything in the universe was a "soupy concoction of plasma compressed into an area smaller than the head of a pin. There was no matter as we think of it now: no iron, no copper, and no oxygen. Just subatomic particles brought together by a crushing force of gravity."[26] No one knows how long the universe stayed in this state. He wrote, "We do know that it was extremely hot, with temperatures exceeding 10 billion degrees, 1,000 times hotter than the center of the sun."[27] Panno wrote, "Eventually, something happened (no one knows what), and that pinhead of unimaginable heat and density suddenly exploded"[28] When reading books on evolution, we can see the varying assumptions made about the conditions at the time of the Big Bang. Some writings state that it was not a Big Bang explosion like a bomb detonating but a rapid expansion from a small point. Panno states that within seconds of the explosion, the extremely high temperature dropped low enough for atomic nuclei to form. I am curious what proof do we have that the temperature was a certain degree? What specific chemical formations happened because the conditions were just right? Panno wrote that after a million years the temperature was low enough for the first element to appear like hydrogen. "The first of these was hydrogen, the simplest of all elements, and the one that gave rise to all the rest."[29] What ever happened to the details? What does the vague language of "gave rise" mean? If we mean produced or created, then tell me the details and steps for this process. Which necessary chemicals came next, and why was is it important for the specific order to occur? Panno then wrote more about the appearance of chemicals in the universe. "…Although the universe was cooling down, it was still hot enough to fuse hydrogen atoms to produce helium. Enough hydrogen and helium were formed in this way to produce all the stars and galaxies"[30] Please, someone tell me how this process worked and which galaxies or optical illusion came first and why? Somewhere between 15,000,000,000 and 10,000,000,000 years, all of this happened and the only information we have about the process is vague language like "gave rise" without an explanation how the rest of the periodic table came about. But somehow we know enough to specify that it was still hot enough to modify hydrogen atoms. Panno then states that 10,000,000,000 years after the Big Bang explosion earth was created as a molten ball of metal and stone thrown off from the sun during the formation of the solar system. Additional materials were added to our planet from asteroids smashing into our planet. The earth freed water due to the high temperature – water vapor from rocks released into atmosphere. This vapor along with dust allowed the earth to cool and it started to rain hundreds of years, about a half billion years has passed.[31] Is there proof of the additional material that was added to earth or the asteroids smashing into earth those many, many years ago?

The First Cell

Panno also wrote that the earth's land was barren and volcanic eruptions were numerous that spewed gases methane CH_4 ammonia NH_1 into atmosphere.[32] Panno mentions that "Modern cells need oxygen to breathe and required four kinds of organic molecules: amino acids (building block for proteins), nucleic acids (building block for DNA and RNA), fats, and sugars. This is a short list, but a long way from methane and ammonia"[33]

Panno mentions that the Urey–Miller experiment showed that some of the basic building blocks for life could have been made in the harsh prebiotic environment. This experiment in 1953 was considered proof of life from non-life but later was weakened because of knowledge of the way modern cells depend upon proteins and nucleic acids, in short DNA and RNA. Panno wrote, "The proteins are used to construct the cell, and a special group of them, called enzymes, control the many chemical reactions that are necessary of the cell to live. DNA is a collection of blueprints, or genes, that store the information to make the protein. One kind of RNA, called messenger RNA (mRNA), serves as an intermediary between the genes and the cell's machinery for synthesizing proteins. Although it is possible for the nucleic acid and proteins to self-assemble it is extremely unlikely that the modern relationship between the three developed spontaneously."[34] The irreducible complexity of the mechanisms presents a serious roadblock to evolution. Because of the extreme complexity of the relationship between proteins and nuclei acid, DNA. The Oxford Dictionary defines DNA as "deoxyribonucleic acid, a self-replicating material which is present in nearly all living organisms as the main constituent of chromosomes, it is the carrier of genetic information."[35] RNA is defined by the Oxford Dictionary as "Ribonucleic acid, a nucleic acid present in all living cells. Its principal role is to act as a messenger carrying instructions from DNA for controlling the synthesis of proteins, although in some viruses RNA rather than DNA carries the genetic information. "[36]

Molecules RNA, DNA Proteins

Panno explains that there are six basic molecules of the cell: Amino acids, Phosphate, Glycerol, Sugars, Fatty acid, Nucleotides. These six basic molecules are used by all cells to construct five essential macromolecules. The five essential macromolecules are Proteins, RNA, DNA and Phospholipid which is defined as a lipid containing a phosphate group in its molecule and Polysaccharide defined as a carbohydrate (e.g. starch, cellulose, or glycogen) whose molecules consist of a number of sugar molecules bonded together.[37] Referring to the molecules of the cell Panno wrote, "All the molecules…are assumed to have formed in the prebiotic oceans, and this was followed by auto-assembly of the macromolecules. Auto assembly of the nucleic acids could have produced polymers…"[38] Wow, the basic molecules formed themselves in the prebiotic soup! I wonder how long it took for these first molecules to form in a beneficial order necessary for life? Do we have valid experimental proof of molecules formed or being present in the prebiotic soup?

Panno wrote that RNA has the ability to replicate itself without the aid of any other molecule. This find is necessary because the molecule that produced the first cell would have to be able to replicate itself and also function as an enzyme. DNA and RNA cannot build molecules by themselves. Proteins cannot duplicate themselves, but they are able to make enzymes. In 1983, Thomas Cech found that RNA molecules were able to have enzymatic activity and was awarded a Nobel prize in chemistry. Of course the idea that the cell responsible for the origin of life started from RNA spread quickly.[39] Panno then wrote, "Eventually, a protein enzyme appeared that could copy RNA into DNA (such an enzyme, called reverse transcriptase, does exist), and when that happened, the cell's machinery approached a modern level of organization: DNA serving as the blueprint, and RNA acting as an intermediary in the process of the protein synthesis. Shifting to a DNA based DNA, as a double-stranded molecule, is more stable than RNA and thus capable of storing information for many genes."[40]

Crick mentions an experiment done by Leslie Orgel and his colleagues that attempted to construct a copying system in a test tube without using protein. The experiment had modest success, but the replication system had some dilemmas. "Even if these difficulties are overcome, the system, though simple, is already somewhat sophisticated. It is, for example, unnaturally pure. It is difficult to imagine how a little pond with just these components, and no others, could have formed on the primitive earth. Nor is it easy to see exactly how the precursors would have arisen.[41] The copying experiment presents an environment that could not have been possible on the earth. The laboratory and test tubes eliminate the outside forces that could make or break the copying system. Crick adds that we do not even know how the precursors came about in the system. There are just too many unknowns, too many questions unanswered to assume the evolution of life as fact. I also wonder about the enzyme that was discovered that showed an ability to copy RNA in to DNA. Do we have proof that the process is sufficient for life origins or do we just say this has happened in this particular case and conclude that it will be the same for life to develop. Panno seems to know that "eventually a protein enzyme appeared" yet the reason why and how are not explored. After such an unknown event the cells machinery reached modernity. Panno explains that even with the RNA as the self-replicator we need to separate the ribozymes and proteins from other chemicals. This is possible because of phospholipids being present in the prebiotic soup. Because of its oily nature it helps the prebiotic soup to form neat bubbles. Of course, we cannot prove that this actually happened during the formation of life on earth. Panno wrote, "Coincidentally, among the organic molecules synthesized in the prebiotic oceans was the oily compound phospholipid."[42] The prebiotic soup is truly amazing – it has everything that scientists need for life at the right time and place. Francis Crick wrote about the prebiotic soup, "Certainly nobody has been able to cook up a primitive soup with water, salts, a few gases and ultraviolet light (or some other energy source) and let it stew away till a neat RNA replicating system arose from it. This failure is not too surprising, since it may have taken nature many millions of years, in many places on the earth's surface, before one happy combination of circumstances produced a system which could both initiate replication and also keep going for some time."[43] I wonder what the probability of such an event would be. Oh I forgot, we cannot come up with (p), so I guess we will have to forgo. Even if RNA copying happened as suggested, Crick wrote, "We have yet to work out how it became coupled to even a primitive form of protein synthesis, although we can begin to make some educated guesses as to how this might have happened."[44] When I think about life origin theories, it seems as though all of the chemicals needed for the beginning of life were coincidentally present. I cannot help but think that without an explanation of how this happened along with the reason why it was necessary to happen in a specific order, we are simply taking what we already know to be necessary for life and building an imagined environment using some scientific facts to fit our theory. No one was present billions of years ago, so we can't ever be sure of what happened. What other chemicals were present? Surely these were not the only molecules. How would they affect each other, and possibly change or destroy the necessary macromolecules of life? How did their presence improve or support life since we seem to know which chemicals were present or do we?

Recently Nature.com reported on their website as story titled DNA has a 521-year half–life. This story, by Matt Kaplan, reports a study of fossils done in New Zealand that clears up how long DNA lasts. When a cell dies, the enzymes break down the bonds between nucleotides that are the backbone of DNA. Paleogeneticist Morten Allentoft used bones of the extinct Moa to calculate the half-life of DNA. The researchers reported that DNA has a half-life of 521 years. After this time, half of the bonds will be broken, then after another 521 years the remaining bonds, if any, would have disappeared. The article reported that Simon Ho, a computational evolutionary biologist, said "This confirms the widely held suspicion that claims of DNA from dinosaurs and ancient insects trapped in amber are incorrect." [45] Examining DNA that is older than about a thousand years is almost useless because the bonds will have broken down. Therefore even if we found DNA that is very old we would not be able to use it to verify our theories.

Panno wrote, "Biologists believe that phospholipids were produced by the storms of ancient earth, forming Earth's first oil slick very close to shore, in relatively calm bays and lagoons" [46] Is this because probably a turbulent prebiotic soup would not make a good home for the peaceful, harmonious origin of life? But tell me how is it that the prebiotic soup is forming in a calm lagoon when the developing world has undergone a major expansion (Big Bag) with tremendous temperature changes over time? One would think that things would still be a little bit unsettled. I think that most bodies of water eventually becomes disturbed and turbulent due to tide or natural disasters.

The Urey Miller experiment showed the possibility that certain chemicals could have been present in the prebiotic earth that are useful for the creation of life. There is a difference between proving the possibility of certain chemicals being present in the harsh prebiotic earth and proving that these chemicals if all available would produce life. The best we could say is that, *if* the prebiotic earth was the way we suggest, some of the basic chemicals were available in the environment to enable chemical reactions which could build some complex macromolecules which are needed to support life. The fact is we cannot prove that life origins happened the way we think it did. It proves that chemicals, and molecules could have been formed which are partly necessary for life. To proceed and say this is what happened is not consistent with our available knowledge. There are too many questions, assumptions and speculations for us to say that this is the way life began. Francis Crick wrote about life beginnings "Whenever and wherever it happened, it started a very long time ago, so long ago that it is extremely difficult to form any realistic idea of such a vast stretch of time."[47]   Francis Crick writing about the experiment Stanley Miller conducted in 1953 including the idea that the earth's atmosphere was much different than it is currently. The assumption that the earth billions of years ago had to be vastly different than it is today is this based upon the knowledge that if oxygen was present, certain molecules would not have formed then life would not have begun. Crick also explains that several atoms (hydrogen, nitrogen, chloride, calcium) are needed for life to have gotten started. There were several other related experiments since done, and Crick mentions that these experiments revealed that if small amounts of oxygen are in the atmosphere, the molecules needed for living systems are *not* formed. When the atmosphere has very little amounts of oxygen, it is called reducing. Crick wrote "In summary, we would like to know the approximate composition of the atmosphere on the earth at a time before life was present, and in particular just how reducing or oxidizing it was. At the moment it seems very difficult to come to any firm conclusion on this matter."[48] Clearly, we see that there is no substantial definitive evidence of the prebiotic environment, we do know that if oxygen was present like today then, life would not have developed. Crick wrote about the concept of abiogenesis, "How likely was it, given a soup of one sort or another, that a system arose spontaneously which could evolve by natural selection? Here we face formidable problems."[49]

Now we can all see that there are many things relating to evolution that we do not know for example, the environment on the prebiotic earth. Scientists begin with the Big Bang and then set up the ideal prebiotic earth with its soupy miracle so that it fits within the boundaries of their theories.

Panno wrote regarding the cell, "The first cells, appearing 3,000,000,000 years ago, quickly evolved into ancestral prokaryotes and, about 2,000,000,000 years ago, gave rise to Archaea, Bacteria, and Eukaryotes, the three major divisions of life in the world. Eukaryotes, in turn, gave rise to plants, animals, protozoans, and fungi."[50] These very important cells are evolving into prokaryotes, which the Oxford Dictionary defines as "a microscopic single–celled organism which has neither a distinct nucleus with a membrane nor other specialized organelles. "[51] Archaea, is defined in the Oxford Dictionary as "microorganisms which are similar to bacteria in size and simplicity of structure but radically different in molecular organization. They are now believed to constitute an ancient group which is intermediate between the bacteria and eukaryotes."[52] Bacteria are "a member of a large group of unicellular microorganisms which have cell walls but lack organelles and an organized nucleus, including some which can cause disease.[53] Finally, eukaryotes are "an organism consisting of a cell or cells in which the genetic material is DNA in the form of chromosomes contained within a distinct nucleus. Eukaryotes include all living organisms other than the eubacteria and archaeabacteria."[54] Now that we are past these definitions, we see that Panno leaves us without explanation as to how and why these cells and chemical reactions occurred the way they did. These cells are according to evolution an early part of human tree.

Panno mentions that "Prokaryotes also developed a method for storing some of the energy that is released when glucose is broken down."[55] How this development happened, we do not know. Panno wrote, "A half–million years after the first prokaryotes appeared, some of them learned to build organic molecules by using energy collected from the sun."[56]

Panno wrote, "The ability to make glucose carriers that are embedded into the cell membrane was probably the most important event leading to the transition from the first cell to the ancestral prokaryotes. Once cells learned this trick, they expanded on it very quickly. Proteins were embedded in the membrane that could detect and import other sugars, such as maltose or lactose."[57] We are writing about a single-celled organism that seems to learn and develop that which is beneficial to itself. After developing these abilities somehow, the single-celled organism improved upon the new found ability. All of this amazing ability of a single-celled organism but, as human beings we are unable to explain how it happened with our large brains, reason and logic. Panno wrote, "Prokaryotes have learned to build and repair themselves with the same molecules that they use for food."[58]

Panno also wrote, "No one knows how DNA came to be the cell's gene bank, but it is possible the enzymatic proteins orchestrated the change themselves."[59] These kind of generalized statements about possible cell behavior do not help us understand how life developed. But we are constantly told that this is true and you must believe what we say. Even though we have not detailed much about the subject.

We do know that life is truly amazing, and as Crick defines life he notes that life and living does not necessarily mean thinking or feeling. Crick wrote that all living things have a molecular build which is the protein family. These proteins can have 2000 atoms that create a three-dimensional form. This three-dimensional structure is very, very important to its operation. Extreme heat will loosen the bond that holds the chain together. The polypeptide chain has the ability to fold itself up in a precise order. The molecular construction is done by the cell stringing together amino acids to make up polypeptide chain in the correct order. Crick wrote, "This is a complicated biochemical process, a molecular assembly line..."[60] The length of the polypeptide chain is about 200-plus amino acids long. A good guess would be there are 20 actual possibilities at each section. The number of possibilities is 10^{260} – which is the number 1 followed by 260 zeroes. These possibilities according to Crick only account for amino acids sequences. Some of the sequences would not fold properly and therefore not produce the desired results.[61] Crick offers more insight into the chemical activities that are glossed over by Panno and others. And of course when we look into the reality of chemical action and reactions we see great complexity which explains why we are mostly provided generalized ideas about the mechanisms of actions and reactions.

Mark Pallen, in his book *The Rough Guide to Evolution*, wrote regarding the origins of life, "Another thing that has to be explained is how the chemistry of life got to be sealed off within the cells from the outside world"[62]. These are things that must be explained before we can say that we know for a fact something is true. When we have to assume or believe something we are approaching a *faith-based system*, not a scientific way of thinking.

Crick further wrote about macromolecules – protein, DNA and RNA. RNA and DNA are able to form structures that are similar to the type found in the double helix. "The process of making a single-stranded RNA copy of a stretch of DNA – called transcription – is relatively straightforward and only requires a rather large protein direct it."[63] Crick wrote, "The cell is thus a minute factory bustling with rapid, organized chemical activity."[64] When talking about a cell, which is duplicated over and over again, we are not discussing some easy to come by chemical combination that can easily happen by chance. The chance or probability for cells to develop life is immeasurable. We cannot even begin to accurately come up with the possibilities because there are just too many variables. Crick mentions that even though our DNA and RNA is mostly universal it is not to be thought of as *simplistic*. "The important point to realize is that in spite of the genetic code being almost universal, the mechanism necessary to embody it is far too complex to have arisen in one blow. It must have evolved from something much simpler. Indeed, the major problem in understanding the origin of life is trying to guess what the simpler system might have been."[65] The major problem with guessing what the simpler system would have been is that a guess is unscientific as believing that God created life. Crick mentions, "In spite of all these uncertainties, it seems possible that in some early stage in the earth's history there was a fair amount of water on its surface, and that in such places it consisted of a weak solution of small organic molecules, many of them not unrelated to the raw materials needed to construct proteins, and nucleic acids, together with various salts washed out of the surrounding rocks."[66] Crick clearly lets us know that there is no simple cut-and-dry theory that exists to explain life origins.

Tyson and Goldsmith in their book wrote about the Urey Miller experiment in 1953 and its significance. The experiment proved "that modestly complex molecules called nucleotides, which provide the key structural element for DNA, the giant molecule that carries instructions for forming new copies of an organism. Even so, a long path remains before life emerges from experimental laboratories. An enormously significant gap, so far unbridged by human experiment or invention, separates the formation of amino acids even if our experiments produced all twenty of them, which they do not and the creation of life."[67] Tyson mentions that the simple molecules that are in living organisms can form quickly in many situations however, life does not. Tyson asks, "How does a collection of molecules, even one primed for life to appear, ever generate life itself?"[68] Here is an evolutionary scientist asking these questions about abiogenesis. This question is key to understanding the depth of complexity involved with what we call life.

The cell is very, very complex, and the RNA molecule for now is believed to be the first replicating cell. The RNA molecule can have about 100 base pairs in specific orders which provide the data on how to interact with the environment which directly relates to the advantage or disadvantage of reproducing. The maximum size of a self-replicating molecule with no error correction is about 100 base pairs maximum. The replication process is affected by mutation errors which will result in the substitution of one base pair for another. Manfred Eigen reported in 1971 that the process of mutation would limit the number of base pairs that a molecule can have. This limitation causes the destruction of information in future molecules. This limitation of the size of self-replicating molecules is problematic because life requires larger molecules to encode genetic information. The living cells use enzymes for repairing and to increase the size of molecules. Some of the larger molecules can have millions of base pairs. The cell uses enzymes for error correcting of the molecules and are much larger than 100 base pairs. Experiments done by Eigen and Schuster concluded that the maximum amount of data encoded in early organisms was severely limited. Modern organisms also have error correction encoded in them and requires much larger data encoding. This fact is referred to as Eigen's Paradox. Joel R. Peck and David Waxman summarized their study in *Is Life Impossible? Information, Sex, and the Origin of Complex Organisms*. They wrote, "The earliest organisms are thought to have had high mutation rates. It has been asserted that these high mutation rates would have severely limited the information content of early genomes. This has led to a well-known 'paradox' because, in contemporary organisms, the mechanisms that suppress mutations are quite complex and a substantial amount of information is required to construct these mechanisms. The paradox arises because it is not clear how efficient error-suppressing mechanisms could have evolved, and thus allowed the evolution of complex organisms, at a time when mutation rates were too high to permit the maintenance of very substantial amounts of information within genomes." [69] A question in the same study was, "How could life that is complex enough to suppress mutation to low levels have evolved while mutation rates were quite high?"[70] There have been a few suggested solutions to the problem. One is that the mutation rate is much less than we have assumed, which would allow longer sequence lengths. This allows for a very basic error correcting mechanism. Wow, that was an easy fix!

NOTES

1. Crick, F. Life Itself: Its Origin and Nature. New York, (NY): Simon and Schuster; 1981. 15 p.

2. Crick, F. Life Itself: Its Origin and Nature. New York, (NY): Simon and Schuster; 1981. 14 p.

3. Crick, F. Life Itself: Its Origin and Nature. New York, (NY): Simon and Schuster; 1981. 15 p.

4. Cowen, R. Simulations back up theory that Universe is a hologram. A ten-dimensional theory of gravity makes the same predictions as standard quantum physics in fewer dimensions. Nature. 2013 Dec 10; doi:10.1038/nature.2013.14328

5. Crick, F. Life Itself: Its Origin and Nature. New York, (NY): Simon and Schuster; 1981, 15 p.

6. Crick, F. Life Itself: Its Origin and Nature. New York, (NY): Simon and Schuster; 1981, 87 p.

7. Crick, F. Life Itself: Its Origin and Nature. New York, (NY): Simon and Schuster; 1981, 88 p.

8. Crick, F. Life Itself: Its Origin and Nature. New York, (NY): Simon and Schuster; 1981, 49 p.

9. Darwin, C. The Origin of Species. Or the preservation of favoured races in the struggle for life [Internet]. London: 2014 Sep 26 [2013 Jan 22] Available from http://www.gutenberg.org/files/1228/1228-h/1228-h.htm

10. Crick, F. Life Itself: Its Origin and Nature. New York, (NY): Simon and Schuster; 1981, 51 p.

11. Crick, F. Life Itself: Its Origin and Nature. New York, (NY): Simon and Schuster; 1981, 88 p.

12. Crick, F. Life Itself: Its Origin and Nature. New York, (NY): Simon and Schuster; 1981, 100 p.

13. Crick, F. Life Itself: Its Origin and Nature. New York, (NY): Simon and Schuster; 1981, 100–101 p.

14. Crick, F. Life Itself: Its Origin and Nature. New York, (NY): Simon and Schuster; 1981,   117 p.

15. Crick, F. Life Itself: Its Origin and Nature. New York, (NY): Simon and Schuster; 1981,   144 p.

16. Crick, F. Life Itself: Its Origin and Nature. New York, (NY): Simon and Schuster; 1981,   145 p.

17. Pallen, MJ. The Rough Guide to Evolution. Rough Guides. London; New York: Penguin Group 2009. 151 p.

18. Ansari, A, Landau, E. Mars Curiosity rover finds life-supporting chemicals [Internet]. CNN Tech; 201 Sep 26 [2013 Dec 15] Available from http://www.cnn.com/2013/ 12/ 10/tech/mars-curiosity-rover/index.html

19. National Research Council of the National Academies. Health Risks from Exposure to Low Levels of Ionizing Radiation: BEIR VII Phase 2. Washington (DC): The National Academies Press; 2006 18–20 p.

20. Shostak, S. Why doesn't NASA just look for life? [Internet]. Huffington Post; 2013 Dec 17 [2013 Aug 24]. Available from http://www.huffingtonpost.com/seth-shostak/why-doesnt-nasa-just-look_b_3808201.html

21. Crick, F. Life Itself: Its Origin and Nature. New York, (NY): Simon and Schuster; 1981, 153 p.

22. Mitton, S. Conflict In The Cosmos Fred Hoyle's Life In Science. Washington, (DC): Joseph Henry Press, 2003. XI p.

23. The Holy Bible. King James Version

24. The Holy Bible. King James Version

25. Panno, J. The cell: evolution of the first Organism (The New Biology Series).   New York (NY): 2005. Facts On File, Incorporated; 1 p.

26. Panno, J. The cell: evolution of the first Organism (The New Biology Series).   New York (NY): 2005. Facts On File, Incorporated; 1 p.

27. Panno, J. The cell: evolution of the first Organism (The New Biology Series).   New York (NY): 2005. Facts On File, Incorporated; 1 p.

28. Panno, J. The cell: evolution of the first Organism (The New Biology Series).   New York (NY): 2005. Facts On File, Incorporated; 1 p.

29. Panno, J. The cell: evolution of the first Organism (The New Biology Series).   New York (NY): 2005. Facts On File, Incorporated; 1 p.

30. Panno, J. The cell: evolution of the first Organism (The New Biology Series).   New York (NY): 2005. Facts On File, Incorporated; 1-2 p.

31. Panno, J. The cell: evolution of the first Organism (The New Biology Series).   New York (NY): 2005. Facts On File, Incorporated; 2 p.

32. Panno, J. The cell: evolution of the first Organism (The New Biology Series).   New York (NY): 2005. Facts On File, Incorporated; 3 p.

33. Panno, J. The cell: evolution of the first Organism (The New Biology Series).   New York (NY): 2005. Facts On File, Incorporated; 4 p.

34. Panno, J. The cell: evolution of the first Organism (The New Biology Series).   New York (NY): 2005. Facts On File, Incorporated; 5 p.

35. Oxford Dictionary online. DNA [Internet]. 2014 Jan 14: Oxford University Press; Available from http://www.oxford dictionaries. com/definition/english/RNA?q=rna

36. Oxford Dictionary online. RNA [Internet]. 2014 Jan 14: Oxford University Press; Available from http://www.oxford dictionaries. com/definition/english/Rna

37. Panno, J. The cell: evolution of the first Organism (The New Biology Series). New York (NY): 2005. Facts On File, Incorporated; 7-9 pp.

38. Panno, J. The cell: evolution of the first Organism (The New Biology Series). New York (NY): 2005. Facts On File, Incorporated; 11 p.

39. Panno, J. The cell: evolution of the first Organism (The New Biology Series). New York (NY): 2005. Facts On File, Incorporated; 11 p.

40. Panno, J. The cell: evolution of the first Organism (The New Biology Series). New York (NY): 2005. Facts On File, Incorporated; 85 p.

41. Crick, F. Life Itself: Its Origin and Nature. New York, (NY): Simon and Schuster; 1981, 85 p.

42. Panno, J. The cell: evolution of the first Organism (The New Biology Series). New York (NY): 2005. Facts On File, Incorporated; 13 p.

43. Crick, F. Life Itself: Its Origin and Nature. New York, (NY): Simon and Schuster; 1981, 85 p.

44. Crick, F. Life Itself: Its Origin and Nature. New York, (NY): Simon and Schuster; 1981. 87 p.

45. Kaplan M. DNA has a 52-year half-life [Internet]. Nature.com; 2013 Dec18 [cited 2012 Oct 10]. Available from

http://www. nature.com/news/dna-has-a-521-year-half-life-1.11555Nature.com

46. Panno, J. The cell: evolution of the first Organism (The New Biology Series).   New York (NY): 2005. Facts On File, Incorporated; 13 p.

47. Crick, F. Life Itself: Its Origin and Nature. New York, (NY): Simon and Schuster; 1981, 19 p.

48. Crick, F. Life Itself: Its Origin and Nature. New York, (NY): Simon and Schuster; 1981, 76 p.

49. Crick, F. Life Itself: Its Origin and Nature. New York, (NY): Simon and Schuster; 1981, 80 p.

50. Panno, J. The cell: evolution of the first Organism (The New Biology Series).   New York (NY): 2005. Facts On File, Incorporated;   xv p.

51. Oxford Dictionary online. Prokaryotes [Internet]. 2014 Jan 14: Oxford University Press; Available from http://www.oxford dictionaries.com/us/definition/american_english/prokaryote?q=prokaryotes

52. Oxford Dictionary online. Archaea [Internet]. 2014 Jan 14: Oxford University Press; Available from http://www.oxford dictionaries.com/us/definition/american_english/archaea

53. Oxford Dictionary online. Bacteria [Internet]. 2014 Jan 14: Oxford University Press; Available from http://www.oxford dictionaries.com/us/definition/american_english/bacteria

54. Oxford Dictionary online. Eukaryote [Internet]. 2014 Jan 14: Oxford University Press; Available from http://www.oxford dictionaries.com/us/definition/american_english/eukaryote?q=eukaryotes

55. Panno, J. The cell: evolution of the first Organism (The New Biology Series).   New York (NY): 2005. Facts On File, Incorporated; 24 p.

56. Panno, J. The cell: evolution of the first Organism (The New Biology Series).   New York (NY): 2005. Facts On File, Incorporated; 24 p.

57. Panno, J. The cell: evolution of the first Organism (The New Biology Series).   New York (NY): 2005. Facts On File, Incorporated; 25 p.

58. Panno, J. The cell: evolution of the first Organism (The New Biology Series).   New York (NY): 2005. Facts On File, Incorporated; 27 p.

59. Panno, J. The cell: evolution of the first Organism (The New Biology Series).   New York (NY): 2005. Facts On File, Incorporated;

60. Crick, F. Life Itself: Its Origin and Nature. New York, (NY): Simon and Schuster; 1981, 51 p.

61. Crick, F. Life Itself: Its Origin and Nature. New York, (NY): Simon and Schuster; 1981, 51 p.

62. Pallen, MJ. The Rough Guide to Evolution. Rough Guides. London; New York: Penguin Group 2009. 154 p.

63. Crick, F. Life Itself: Its Origin and Nature. New York, (NY): Simon and Schuster; 1981, 70 p.

64. Crick, F. Life Itself: Its Origin and Nature. New York, (NY): Simon and Schuster; 1981, 70 p.

65. Crick, F. Life Itself: Its Origin and Nature. New York, (NY): Simon and Schuster; 1981, 71 p.

66. Crick, F. Life Itself: Its Origin and Nature. New York, (NY): Simon and Schuster; 1981, 80 p.

67. Tyson, ND, Goldsmith, D. Origins fourteen billion years of cosmic evolution. New York–London: W.W. Norton & Company; 2005. 243 p.

68. Tyson, ND, Goldsmith, D. Origins fourteen billion years of cosmic evolution. New York–London: W.W. Norton & Company; 2005. 243 p.

69. Peck JR, Waxman, D. Is Life Impossible? Information, Sex, and the Origin of Complex Organisms. Evolution. 2010 Nov; 64(11): 3300-3009 doi:10.1111/j.1558-5646.2010.01074.x

70. Peck JR, Waxman, D. Is Life Impossible? Information, Sex, and the Origin of Complex Organisms. Evolution. 2010 Nov; 64(11): 3300-3009 doi:10.1111/j.1558-5646.2010.01074.x

Chapter 7
Evolution as fact

I had the opportunity to speak at a symposium on the Bible and evolution in Brooklyn, NY. I had to use Richard Dawkins' book, *The Greatest Show on Earth* in a mock debate. The purpose was to educate people about some of the issues surrounding evolution and to give them a well-rounded picture of both sides. I was to represent a person that did not accept the theory of evolution. I had to give reasons why I disagreed with some of Dawkins' points in the book. Dawkins referred to people that do not agree with evolutionary theory as "history deniers" and also associated them with people that deny the Holocaust ("Holocaust deniers"). This is ridiculous and insensitive. There is proof of the Holocaust and the evil done to Jewish people. The Holocaust did happen, just like slavery happened, and we must never forget or we risk repeating history. Dawkins wrote, "The history-deniers themselves are among those that I am trying to reach in this book. But, perhaps more importantly, I aspire to arm those who are not history-deniers but know some – perhaps members of their own family or church and find themselves inadequately prepared to argue the case."[1] After I read the book, I realized that Dawkins like many others did not offer enough details but only pictures of species believed to have gone through macroevolution. So as a so-called history-denier, I am no closer to accepting all evolution theory than before I read his book. Dawkins focuses on what he called fossil proof, even though he later downgrades the same evidence as being unimportant and not necessary because evolution could and does stand on its own. However, he spends a significant percentage of his book going over various pictures of organisms that have been examined and supposedly support evolution. The believers in evolution that he is trying to reach will simply have the generalized statements that do not go into the details of evolution to see if it all makes sense. I have great concern over Dawkins writing that "Evolution is a fact. Beyond reasonable doubt, beyond serious doubt, beyond sane, informed, intelligent doubt, beyond doubt evolution is a fact. The evidence for evolution is at least as strong as the evidence for the Holocaust, even allowing for eye witnesses to the Holocaust."[2] I think that the Holocaust is way too serious an issue and deserves the proper respect and should not be discussed along with evolutionary theory. Here, I will only deal with Dawkins statements about evolution. This is the first time I ever read a book by someone that stated evolution is a fact. Unfortunately, I think that this is only the beginning and many others will be emboldened to follow. The general public will be even more confused, and evolutionary ideas will spread even more. This is why this book and others like it need to be produced so that we will be reminded of some of the details, assumptions and problems with evolutionary thought. Most scientists will not even know that this book exists because it has not been christened or approved by one of their academic houses of worship. Forgetting that before there was a university and scientific degrees, there was common sense. Dawkins repeats, "Evolution is a fact, and this book will demonstrate it."[3] Unfortunately, the public is used to quick sound bites and cliché's. Dawkins also wrote, "Evolution is a fact in the same sense as it is a fact that Paris is in the Northern Hemisphere."[4]

Dawkins proceeds to tell us about a hypothetical crime story with a detective and a spy camera. The story is about a crime that has been committed and the detectives (scientists) that come on the scene after the crime and put together the story of what happened. Dawkins wrote "the murderer's actions have vanished into the past. The detective has no hope of witnessing the actual crime with their own eyes. In any case the gorilla-suit experiment and others similar to it have taught us to mistrust our own eyes. What the detective does have is traces that remain, and there is a great deal to trust there. There are footprints, fingerprints and nowadays DNA fingerprints too, bloodstains, letters, diaries."[5] Dawkins is using this example to show us that we cannot trust what we see. Of course, he is specifically trying to convince us not to expect a fossil record and remember that we cannot trust our eyes. Dawkins, however, used several pictures in his book to show us the "truth" and "proof" of evolution. Confusing?

Dawkins in chapter three wrote, "This chapter embarks on a step-by-step seduction of the minds as we pass from the familiar territory of dog breeding and artificial selection to Darwin's giant discovery of natural selection, via some colorful intermediate stages."[6] We have already clarified earlier in this book that Charles Darwin was not able to look at the miniscule. However, today, as biologist Michael Behe wrote, we can no longer only look at organisms as a whole to determine evolution because the minutia tells the compelling story. Richard Dawkins is right about his book when he mentions that he is trying to seduce the mind by showing pictures and presenting general evolutionary ideas. Behe wrote in *Darwin's Black Box* about Darwinian evolution that "To many, its triumph seems complete. But the real work of life does not happen at the level of the whole animal or organ; the most important parts of living things are too small to be seen. Life is in the details, and it is molecules that handle life's details. Darwin's idea might explain horse hoofs, but can it explain life's foundation?"[7] Behe adds, "Because the popular media likes to publish exciting stories, and because some scientists enjoy speculating about how far their discoveries might go, it has been difficult for the public to separate fact from conjecture. To find the real evidence you have to dig into journals and books published by the scientific community itself. The scientific literature reports experiments firsthand, and the reports are generally free of the flight of fancy that make their way into the spinoffs that follow."[8] I have tried to access to some of these databases to study from home, but the cost is outrageous for something like Jstor. The general public hardly reads, so why would they pay hundreds or thousands of dollars for access to Jstor or other academic databases. One can go to a local university but that is old school and inconvenient.

Behe explained that we have examples of what can be called microevolution but we do not have solid examples of macroevolution, and the so-called universal common ancestor (UCA) that everything springs from. He wrote, "Roughly speaking, *microevolution* describes changes that can be made in one or a few small jumps, whereas *macroevolution* describes changes that appear to require large jumps. Behe explains that microevolution and Darwin's theory has triumphed. He wrote

"But it is at the level of macroevolution – of large jumps – that the theory evokes skepticism."[9]

Dawkins wrote about the various dating methods used to record time periods like using tree rings, radioactive clocks, and carbon dating. He then wrote about the way we recognize geological layers and their order. Dawkins notes, "...because the layers are so recognizable, you can work out their relative ages by daisychaining and jigsawing your way around the world."[10] Why would Dawkins use these words to describe an important geological fact? I will tell you what I think: Even though we can use the geological layers in science, from what I have read the geological column as in textbooks are shown it to be 100 to 200 miles deep, but we can only find about ten to twenty percent of the actual miles in reality, which is a big difference. Richard Dawkins continues, "So, long before we knew how old fossils were, we knew the order in which they were laid down, or at least the order in which the named sediments were laid down..."[11] He adds, "The named strata are usually identified by the fossils they contain. And we are going to use the ordering of the fossils as evidence for evolution!" Dawkins then asks "Is that in danger of turning into a circular argument?"[12] Dawkins goes on to assert that the argument is not circular.

Dawkins also wrote about the missing links by asking *What do you mean missing?* He goes back to the story about the detective coming to the crime scene after the crime and piecing together the events. Dawkins wrote, "The fossil record, like the spy camera in the murder story, is a bonus, something that we had no right to expect as a matter of entitlement. There is more than enough evidence for the fact of evolution in the comparative study of modern species (chapter 10) and their geographical distribution (chapter 9)"[13] Dawkins goes on, "We don't need fossils – the case for evolution is watertight without them; so it is paradoxical to use gaps in the fossil record as though they were evidence against evolution. We are, as I say, lucky to have fossils at all."[14] Is it really unreasonable to expect to find some convincing evidence somewhere in the entire world that would provide a continuous line of evolution that has occurred over many, many millions of years? Dawkins later admits, "Yes, there are gaps, where there are no fossils at all, and that is only to be expected…"[15] Dawkins surprisingly mentions the sudden appearance of the "great animal phyla- the main divisions of the animal world"[16] in the fossil record. There is no proof of these animals in rocks older than the Cambrian. These animals appear in an advanced state of evolution the very first time we see them. Dawkins goes on to write about fossils in human evolution: "I wish we really did have a complete and unbroken trail of fossils, a cinematic record of all evolutionary change as it happened."[17] Ok, Dawkins, now even you would have liked to have the ultimate proof of the fossil record to put to bed the questions about evolution. Well, we wish that you had them, too; that's not so unreasonable is it? We can all wish, can't we?

Dawkins also wrote about the foundation of life. Chapter 8 starts off telling a secondhand story about a Professor Haldane being questioned by a lady after one of his lectures on evolution. She asked how can given even a billion years evolution go from a single cell to a multi-cell human body? The professor responded "But madam, you did it yourself, and it only took you nine months."[18] This type of cute response, a joke or not, oversimplifies the reality of the complications involved in DNA, RNA and cell operation. For a scientist to even attempt to explain DNA and RNA in detail as it relates to their theory of evolution would be impossible. Dawkins amazingly admits, "The detail of how the laws of chemistry determine the tertiary structure of a protein are not yet fully understood: chemists can't yet deduce, in all cases, how a given sequence of amino acids will coil up."[19] This is very important in chemistry because the way that the amino acid coils up is extremely important if the desired life/function is to be produced.

A response to Dawkins' colorful pictures of organisms that are supposed to have evolved, Michael Behe states "…it is no longer enough for an evolutionary explanation of that power to consider only the anatomical structure of whole eyes, as Darwin did in the nineteenth century (and as popularizers of evolution today). Each of the anatomical steps and structures that Darwin thought were so simple actually involves staggeringly complicated biochemical processes that cannot be papered over with rhetoric."[20] As Crick mentioned in his book, the evolution of life as we know it today must have become the way it is through small steps advancing from a simpler organism. However, as we dig into the minute – the building blocks of all life – the complexity does not get simpler but even more complex. There appears to be no lessening of complexity when it comes to human life. Behe wrote, "Anatomy is, quite simply, irrelevant to the question of whether evolution could take place on the molecular level. So is the fossil record."[21] This is what I call going to the bottom line.

Dawkins mimics the book of Ecclesiastes chapter 12 verse 13, "Let us hear the conclusion of the whole matter: Fear God, and keep his commandments: for this is the whole duty of man."[22] He writes, "let us hear the conclusion of the whole matter. There is no overall plan of development, no blueprint, no architect's plan, no architect. The development of the embryo, and ultimately of the adult, is achieved by local rules that apply to molecules, especially protein molecules, within the cells and in the cell membranes, interacting with other such molecules. Again the rules are all local, local, local."[23] Thank you, Dawkins, for reminding us that life came by chance from non-life and not from a superintelligent being, supreme designer or God.

Irreducible complexity

We simply cannot get away from this fact of irreducible complexity though we have touched a little on the topic earlier. This is really evident when we stop carelessly glossing over the fine details. We can no longer proceed in the imaginative. The idea of irreducible complexity suggests that certain biological systems are too complicated to have evolved from a simpler form via natural selection. Behe is credited as the originator of this term. He stated that irreducible systems are made of several well-matched, interacting parts that provide functionality, and if any part is removed the system will stop functioning.[24] Charles Darwin wrote about it himself, but I guess some scientists would like to ignore his point: "If it could be demonstrated that any complex organ existed which could not possibly have been formed by numerous, successive, slight modifications, my theory would absolutely break down."[25] In the book *Guide to Evolution,* Pallen wrote "Throughout the *Origin*, Darwin constantly raises objections to his own hypotheses, only to demolish them"[26] Really, I do not see where he has demolished all of the concepts that he raised especially those about complex organisms. The statement is not entirely true, as Darwin could not look at the complexity of the cell the way we can today.

There was a YouTube video that dealt with the book *Darwin's Black Box* and some of the details. The debate was titled *Darwin's Black Box* in which Behe is talking about the cell and its irreducible complexity. The fact is that all scientific ideas related to life origins today must fit within the boundaries of evolution to be considered worthy of discussion. In the video when Keith Fox was asked what if we proved that evolution was falsified, would the evolutionists admit that the theory failed. Of course, the question was not answered, but the response given goes back to what scientists cannot do and so on. Michael Behe was asked by the debater if at some point we can prove that the things you say are irreducible really are not and can happen by Darwinian evolution, where does that leave intelligent design and your God creator? Behe replied that if it was proved and indeed was not irreducible "I would be wrong." Behe gave a direct answer whereas Keith Fox the evolutionist dodged the point. Behe then asked what happens if the reverse occurred and things were irreducible and there was no Darwinian evolutionary outcome. Would Keith Fox say then that Darwinian evolution had been falsified, and if not what was his criteria for falsifying Darwinian evolution? Fox replied but did not directly answer the question, "Because we can't go back in time to know what those evolution pressures were that caused the bacteria flaggelan or its originator."[27] Basically Keith Fox did not answer the question. Michael Behe then suggested that we can experiment. Keith Fox then said, "We can do experiments based on what we hypothesized might be those conditions." The response given by Fox gives no clear answer to the question asked. The moderator then asked, "Is there a problem here then, Keith?" The moderator was referring to the idea that intelligent design does not give us an explanation of life, it just states that God did it. Similarly evolutionary thought is assumed from the outset and everything has to fall under its title/category of evolution. Evolution is the one thing that is not questioned in science. The moderator said "Well surely we need to have a skeptical attitude to evolution itself not just assume the thing we don't understand will automatically fall under its umbrella." Keith Fox responded that scientists are skeptical and they want to know why, if not satisfied. The moderator then asked, "But they wouldn't go outside of evolution?"[28] Fox said, "They probably wouldn't, but it is within the context of evolution can we put together a realistic scenario that's not just a just-so story."[29] This is an important point because the scientific community does not have an open door for any other ideas except Darwinian evolution. Within Darwinian thought one can suggest and put forth various ideas. What is interesting is that Behe answered the question directly. But the other debater, Keith Fox, did not give a direct answer, he simply diverted the conversation to what cannot be proven because we cannot go back in time. Then he also says that anything solutions concerning life origins has to be within the bounds of evolutionary thought. It is these strict boundaries that refuse to acknowledge any ideas or explanations that are not centered in evolutionary thought. This intentional redirection away from other ideas that are not under the title of evolution is binding of free thought, which hinders true scientific discovery. For example, when we think back on dark matter we see that even though we cannot define what dark matter is, we see its effects. Dark matter does not fit our current understanding of matter, so maybe it is something new that does not fit our scientific rules. I do not think that thinking outside of the evolutionary box is the end of the world. In fact, it causes us to examine the phenomenon more closely. Sometimes this close examination results in understanding that the phenomenon is not a miracle but can be explained. Part of the challenge is realizing that we do not have the tools to understand a phenomenon nor the patience and humility to admit we just do not know yet.

NOTES

1. Dawkins, R. The Greatest Show on Earth: The Evidence for Evolution. New York, (NY): Free Press; 2009. 8 p.

2. Dawkins, R. The Greatest Show on Earth: The Evidence for Evolution. New York, (NY): Free Press; 2009. 8 p.

3. Dawkins, R. The Greatest Show on Earth: The Evidence for Evolution. New York, (NY): Free Press; 2009. 9 p.

4. Dawkins, R. The Greatest Show on Earth: The Evidence for Evolution. New York, (NY): Free Press; 2009. 10 p.

5. Dawkins, R. The Greatest Show on Earth: The Evidence for Evolution. New York, (NY): Free Press; 2009. 16 p.

6. Dawkins, R. The Greatest Show on Earth: The Evidence for Evolution. New York, (NY): Free Press; 2009. 46 p.

7. Behe, MJ . Darwin's Black Box: The biochemical challenge to evolution. New York (NY) : The Free Press; 1996. 5 p.

8. Behe MJ . Darwin's Black Box: the biochemical challenge to evolution. New York, (NY): The Free Press; 1996. 5 p.

9. Behe MJ . Darwin's Black Box: the biochemical challenge to evolution. New York, (NY): The Free Press; 1996. 15 p.

10. Dawkins, R. The Greatest Show on Earth: The Evidence for Evolution. New York, (NY): Free Press; 2009. 99 p.

11. Dawkins, R. The Greatest Show on Earth: The Evidence for Evolution. New York, (NY): Free Press; 2009. 99 p.

12. Dawkins, R. The Greatest Show on Earth: The Evidence for Evolution. New York, (NY): Free Press; 2009. 99 p.

13. Dawkins, R. The Greatest Show on Earth: The Evidence for Evolution. New York, (NY): Free Press; 2009. 146 p.

14. Dawkins, R. The Greatest Show on Earth: The Evidence for Evolution. New York, (NY): Free Press; 2009. 146 p.

15. Dawkins, R. The Greatest Show on Earth: The Evidence for Evolution. New York, (NY): Free Press; 2009. 147 p.

16. Dawkins, R. The Greatest Show on Earth: The Evidence for Evolution. New York, (NY): Free Press; 2009. 147 p.

17. Dawkins, R. The Greatest Show on Earth: The Evidence for Evolution. New York, (NY): Free Press; 2009. p.

18. Dawkins, R. The Greatest Show on Earth: The Evidence for Evolution. New York, (NY): Free Press; 2009. 211 p.

19. Dawkins, R. The Greatest Show on Earth: The Evidence for Evolution. New York, (NY): Free Press; 2009. 237 p.

20. Behe, MJ . Darwin's Black Box: The biochemical challenge to evolution. New York (NY) : The Free Press 1996. 22 p.

21. Behe, MJ . Darwin's Black Box: The biochemical challenge to evolution. New York (NY) :The Free Press 1996. 22 p.

22. The Holy Bible. King James Version.

23. Dawkins, R. The Greatest Show on Earth: The Evidence for Evolution. New York, (NY): Free Press; 2009. 247 p.

24. Behe, MJ . Darwin's Black Box: The biochemical challenge to evolution. New York (NY) : The Free Press 1996. 42-46 p.

25. Darwin, C. The Origin of Species. Or the preservation of favoured races in the struggle for life [Internet]. London: 2014 Sep 26 [2013 Jan 22] Available from http://www.gutenberg.org/files/1228/1228-h/1228-h.htm

26. Pallen, MJ. The Rough Guide to Evolution. Rough Guides. London; New York: Penguin Group 2009. 49 p.

27. Darwin's Black Box: Michael Behe & Keith Fox debate intelligent design [Internet]. 2014 Sep 25, [cited 2010 Oct 23]. Available from http://www.youtube.com/watch?v=luRGzVrr2Cs

28. Darwin's Black Box: Michael Behe & Keith Fox debate intelligent design [Internet]. 2014 Sep 25, [cited 2010 Oct 23]. Available from http://www.youtube.com/watch?v=luRGzVrr2Cs

29. Darwin's Black Box: Michael Behe & Keith Fox debate intelligent design [Internet]. 2014 Sep 25, [cited 2010 Oct 23]. Available from http://www.youtube.com/watch?v=luRGzVrr2Cs

Chapter 8
Mutation of Genes (Natural Selection)

The University of California Museum of Paleontology has a website that deals with mutations and evolution. The website stated that "Mutations can be beneficial, neutral, or harmful for the organism,.."[1] When a mutation occurs in genes there is a change in the DNA. The website states that "A single mutation can have a large effect, but in many cases, evolutionary change is based on the accumulation of many mutations."[2] Tom Strachan and Andrew P. Read in their book *Human Molecular Genetics* wrote, "Mutations are the raw fuel that drives evolution, but they can also be pathogenic. They can be the direct cause of a phenotypic abnormality or they can result in increased susceptibility to disease. The usually low level of mutation may therefore be viewed as a balance between permitting occasional evolutionary novelty at the expense of causing disease or death in a proportion of the members of a species. Normally, most mutations arise as copying errors during DNA replication because DNA polymerases, like all enzymes, are error–prone."[3] I then looked for the mutations that we know about, here are the results. "One after another, genes for important disorders such as Duchenne, muscular dystrophy-gene Dystrophin, cystic fibrosis gene CFTR, Branchio-oto-renal syndrome (BOR) gene EYA1, Treacher Collins syndrome gene TCOF1, Sickle cell diseases, Huntington disease, adult polycystic kidney disease, colorectal cancer, breast cancer, etc..."[4] Using positional cloning, disease genes can be identified using only their approximate chromosomal location. Cancer, the mutation that we all know about and fear having, is a product of genes. "Because cancers are the inevitable end-result of natural selection among the cells of an organism, rather than the result of a specific disease process, cancers of a given type do not all have mutations in a standard set of genes".[5] The website www.cancer.gov mentions that a healthy body is due to the actions of thousands of proteins in harmony and functioning properly. [6] The function of the proteins are directly related to the gene "and each properly functioning protein is the product of an intact gene. Many, if not most, diseases have their roots in our genes. More than 4,000 diseases stem from altered genes inherited from one's mother and/or father."[7] They also stated that "All cancer is genetic, in that it is triggered by altered genes. Genes that control the orderly replication of cells become damaged, allowing the cells to reproduce without restraint."[8] The reason why I have mentioned the point is that most mutations that we know about today are mutations that cause significant harm or even death to the organism.

NOTES

1. The University of California Museum of Paleontology. Welcome to understanding evolution for teachers. [Internet]. 2014 Sep 26 . 2014 Available from http://evolution.berkeley.edu/evosite/evo101/IIIC1Mutations.shtml

2. The University of California Museum of Paleontology. Welcome to understanding evolution for teachers. [Internet]. 2014 Sep 26 . 2014 Available from http://evolution.berkeley.edu/evosite/evo101/IIIC1Mutationsp2.shtml

3. Strachan T, Read AP. Instability of the human genome: mutation and DNA repair. Human Molecular Genetics. 2nd edition. New York: Wiley–Liss; 1999. (Collins, 1995). Chapter 9.1 Available from: http://www.ncbi.nlm.nih.gov/books/NBK7566//NBK7566/ accessed 11-13-13 9:16 a.m

4. Strachan T, Read AP. Instability of the human genome: mutation and DNA repair. Human Molecular Genetics. 2nd edition. New York: Wiley–Liss; 1999. (Collins, 1995). Chapter 15.3 Available from: http://www.ncbi.nlm.nih.gov/books/NBK7566/NBK7566/ accessed 11-13-13 9:16 a.m

5. Strachan T, Read AP. Instability of the human genome: mutation and DNA repair. Human Molecular Genetics. 2nd edition. New York: Wiley–Liss; 1999. (Collins, 1995). Chapter 18.8 Available from: http://www.ncbi.nlm.nih.gov/books/NBK7566//NBK7566/ accessed 11-13-13 9:16 a.m

6. National Cancer Institute at the National Institute of Health. Slide 8 Gene mutation and disease [Internet]. Bethesda (MD): 2014, Sep 26 [cited 2005 Jan 28]. Available from http://www.cancer .gov/cancertopics/understandingcancer /genetesting /page8

7. National Cancer Institute at the National Institute of Health. Slide 8 Gene mutation and disease [Internet]. Bethesda (MD): 2014, Sep 26 [cited 2005 Jan 28]. Available from http://www.cancer.gov/cancertopics/understandingcancer/genetesting/page8

8. National Cancer Institute at the National Institute of Health. Slide 8 Gene mutation and disease [Internet]. Bethesda (MD): 2014, Sep 26 [cited 2005 Jan 28]. Available from http://www.cancer.gov/cancertopics/understandingcancer/genetesting/page8

Chapter 9
Final thoughts

Being a Christian or believing in God does not automatically mean that a person is anti-science. Yet in the debate between Ken Ham and Bill Nye on February 4, 2014, on the topic of whether or not creationism is a viable world view, Nye repeatedly focused on the need to have scientists in America if we are to remain a leading country and so equated creationism/intelligent design with the absence of scientific learning. This argument is simply not truthful. I don't know any Christians that will not go to the doctor or take natural holistic medicine because they don't believe in science. When we have a headache, we may take aspirin to relieve the pain. Going to the herbal store and picking up ginger or vinegar or tea leaves is definitely related to science. Thank God for what has been revealed through science.

Christians do believe in science, however, we have a higher authority that we believe should be the final say-so in matters. How is this, you might say? For example, let's say you have a sick child and the doctor's say that we think that your child must have this or that procedure. I would first listen carefully to what the doctor has said, research the subject, talk to other medical professionals. Finally, I would begin to pray and ask God for guidance. God influences and or directs the final decision. There have been times when I asked the doctor politely for more time to consider the options, and they had no problem with the request. Proverbs 3:6 states that we should acknowledge God in all our ways and He shall direct our paths. Sometimes our healing path is via doctors and science, sometimes it is through divine intervention. I know and have heard of persons who have decided to depend solely on God for their miraculous healing and have died as well as persons that God has healed from their infirmity. The late Bishop Ithiel Clemmons said that God can heal you through the doctors, through death, and through a miracle. The final results are totally up to God.

I do not think that science and the Bible are incompatible. I think that science has its place. Perhaps religion/creationism should have never been part of the school curriculum. However, I do not think that evolution should be taught in schools either. Just like we cannot prove that there is or isn't a God that created life, we also cannot prove that evolution is the correct understanding of life origins. Sure, we do have some scientific theories that make a good attempt at explaining possible origins of life but that is where it ends. We cannot proceed in saying that evolution is a fact without understanding the details of the theory. If there is any area that is not clearly understood in our theories we need to continue to research and learn. However, we should be very, very careful what we teach our children. We cannot honestly write like Richard Dawkins that evolution is a fact; that would be careless.

In this book I wrote about: (a) the elusive dark matter; (b) the Big Bang and problems explaining life origins; (c) probability problems; (d) the Cambrian Explosion; (e) stars that might have planets; (f) extreme complexity shown at the DNA level; (g) the primitive soup insufficient to produce life; and (h) speculation on life origins. If any of the questions or points that I brought out in this book remains ambiguous, unanswered or mysterious, scientists must continue to develop and go back to the drawing board to confirm the truth of theories. Not doing so would be approaching the same type of "blind faith" of which some scientists accuse creationists. Remember, Christians do not claim to be scientists. Scientists make the claim that they have come to their conclusions using the scientific method.

www.ingramcontent.com/pod-product-compliance
Lightning Source LLC
Chambersburg PA
CBHW022001170526
45157CB00003B/1087